1 MONTH OF
FREE
READING

at
www.ForgottenBooks.com

By purchasing this book you are eligible for one month membership to ForgottenBooks.com, giving you unlimited access to our entire collection of over 1,000,000 titles via our web site and mobile apps.

To claim your free month visit:
www.forgottenbooks.com/free894291

ISBN 978-0-266-82021-5
PIBN 10894291

cal Microreproductions / Institut canadien de microreproductions historiques

1997

Technical and Bibliographic Notes / Notes techniqu

The Institute has attempted to obtain the best original copy available for filming. Features of this copy which may be bibliographically unique, which may alter any of the images in the reproduction, or which may significantly change the usual method of filming are checked below.

L'Institut a micro été possible de s plaire qui sont p ographique, qui p ou qui peuvent e de normale de filr

☑ Coloured covers /
Couverture de couleur

☐ Covers damaged /
Couverture endommagée

☐ Covers restored and/or laminated /
Couverture restaurée et/ou pelliculée

☐ Cover title missing / Le titre de couverture manque

☑ Coloured maps / Cartes géographiques en couleur

☐ Coloured ink (i.e. other than blue or black) /
Encre de couleur (i.e. autre que bleue ou noire)

☐ Coloured plates and/or illustrations /
Planches et/ou illustrations en couleur

☐ Bound with other material /
Relié avec d'autres documents

☐ Only edition available /
Seule édition disponible

☐ Tight binding may cause shadows or distortion along interior margin / La reliure serrée peut causer de l'ombre ou de la distorsion le long de la marge intérieure.

☐ Blank leaves added during restorations may appear within the text. Whenever possible, these have been omitted from filming / Il se peut que certaines pages blanches ajoutées lors d'une restauration apparaissent dans le texte, mais, lorsque cela était possible, ces pages n'ont pas été filmées.

☐ Additional comments /
Commentaires su lémentaires:

☐ Coloured pa

☐ Pages dam

☐ Pages resto
Pages resta

☑ Pages disco
Pages déco

☐ Pages detac

☑ Showthroug

☐ Quality of pr
Qualité inég

☐ Includes sup
Comprend c

☐ Pages whol
tissues, etc.
possible i
partiellemer
pelure, etc.,
obtenir la m

☐ Opposing
discolouratic
possible ima
colorations
filmées deu
possible.

2　　**3**　　　　　　　　**1**

2

MICROCOPY RESOLUTION TEST CHART

(ANSI and ISO TEST CHART No. 2)

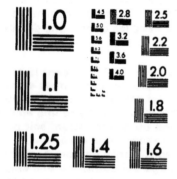

APPLIED IMAGE Inc

1653 East Main Street
Rochester, New York   14609   USA
(716) 482 - 0300 — Phone
(716) 288 - 5989 — Fax

CANADA
DEPARTMENT OF MINES
HON. MARTIN BURRELL, MINISTER; R. G. McCONNELL, DEPUTY MINISTER.

GEOLOGICAL SURVEY
WILLIAM McINNES, DIRECTING GEOLOGIST.

MEMOIR 108

No 92, GEOLOGICAL SERIES

The Mackenzie River Basin

BY

Charles Camsell and Wyatt Malcolm

OTTAWA
J. DE LABROQUERIE TACHÉ
PRINTER TO THE KING'S MOST EXCELLENT MAJESTY
1919

No. 1713

Plate I.

Alexandra falls on Hay river; height 105 feet. (Page 29.)

CANADA

DEPARTMENT OF MINES

HON. MARTIN BURRELL, MINISTER: R. G. McCONNELL, DEPUTY MINISTER.

GEOLOGICAL SURVEY

WILLIAM McINNES, DIRECTING GEOLOGIST.

MEMOIR 108

No 92, GEOLOGICAL SERIES

The Mackenzie River Basin

BY

Charles Camsell and Wyatt Malcolm

OTTAWA
J. DE LABROQUERIE TACHÉ
PRINTER TO THE KING'S MOST EXCELLEN. MAJESTY
1919

No. 1713

57048—1

CONTENTS.

Illustrations.

The Mackenzie River Basin.

GENERAL STATEMENT.

The basin of Mackenzie river and its tributaries has received considerable attention in recent years. It is a section of country through which settlement is gradually extending northward, and through which agricultural development will probably be limited in its northward movement by climatic conditions only. As settlement progresses transportation lines are opened up, and increased interest is taken in all the economic resources of the region. Information about this region is very limited, as the explorations that have been carried on since Alexander Mackenzie descended in 1789 the great river that bears his name, have been confined almost wholly to the principal water routes. It is the purpose of this compilation to present a concise statement of the present knowledge of its economic possibilities, particular attention being given to the geological features and possible mineral resources. To this compilation Charles Camsell, one of the writers, contributes much information acquired through long residence and wide exploration throughout the Mackenzie River basin.

LOCATION AND AREA.

The basin of Mackenzie river occupies the northwestern part of the continent of North America and includes within its area the northern parts of the provinces of British Columbia, Alberta, and Saskatchewan, the western part of the Northwest Territories, and parts of southeastern and northeastern Yukon. It crosses 16 degrees of latitude between the 53rd and the 69th parallels, and 36 degrees of longitude between the 104th and the 140th meridians.

The greater part of the basin lies in what is known as the Great Plains of North America, but a considerable portion is also included within the Laurentian Plateau region on the east and about an equal area within the Cordilleran region on the west.

The total area of the Mackenzie basin is about 682,000 square miles or somewhat more than one-fifth of the area of Canada exclusive of the islands in the Arctic ocean. Water, as lakes, covers a very large part of this area, and although many large lakes of the region are still unexplored and unmapped it is estimated that the water area cannot be less than about 40,000 square miles.

The total length of the river from the headwaters of Finlay river to the Arctic ocean has not yet been accurately measured, but it is estimated to be about 2,525 miles.

The Mackenzie River system ranks as one of the eight great river systems of the earth. On the North American continent it is exceeded in length and drainage area only by the Mississippi river. It has a slightly greater length and larger area than the St. Lawrence.

The following table of lengths and areas, taken mainly from the Atlas of Canada, will serve to give an idea of the importance of the Mackenzie river relative to the other large river systems of the continent of North America.

	Length, miles.	Area, sq. miles.
Mississippi (to head of Missouri)	4,220	1,244,000
Mackenzie (to head of Finlay)	2,525	682,000
St. Lawrence (from Pt. de Monts to head of St. Louis)	1,900	498,500
Yukon (to head of Nisutlin)	1,765	330,000
Nelson (to head of Bow)	1,660	370,800
Colorado (to head of Green)	2,200	225,000
Columbia	1,150	250,000

HISTORY.

The earliest explorations in the basin of the Mackenzie were made in the latter part of the eighteenth century. Samuel Hearne of the Hudson's Bay Company set out from Fort Prince of Wales at the mouth of Churchill river in December 1770, on a journey to Coppermine river to investigate the copper deposits of that region. He reached the river and followed it to its mouth. On his return he took a more westerly course than that followed on his northward journey, crossed Great Slave lake, and ascended Slave river some distance. From this river he travelled eastward and arrived at Fort Prince of Wales at the end of June, 1772.

The first white man to appear on Athabaska river was Peter Pond. His route to the Athabaska lay by way of Churchill river, Ile à la Crosse lake, Buffalo lake, Methye river, Methye lake, Methye portage, and Clearwater river, a route that was used for many years as the readiest means of access to the Mackenzie basin. In 1778 he descended the Athabaska and founded a fort known as the "Old Establishment" about 30 miles above its mouth. He was also, evidently, the first white man to reach the shores of lake Athabaska. In 1786 Pond sent Laurent Leroux and Cuthbert Grant to build a fort on Great Slave lake; these being the first white men, with the exception of Samuel Hearne, to visit this lake. The post was built at the eastern mouth of Slave river.[1]

Philip Turner carried on explorations in this part of the country in 1790-92 and made careful surveys of Athabaska river below the mouth of Clearwater, of Athabaska lake, of Slave river, and of a small portion of the south shore of Great Slave lake.[2]

In 1788, Fort Chipewyan was built on the south shore of lake Athabaska on what is now known as Old Fort point. From this point Alexander Mackenzie set out in the spring of 1789 to explore the river that drains Great Slave lake. In spite of repeated warnings from natives as to the danger of the undertaking he pushed his way with indomitable energy to the mouth of the river and returned to Fort Chipewyan after an absence of only one hundred and two days, accomplishing one of the "most remarkable exploits in the history of inland discovery, whether regarded in the light of the results achieved, or of the time taken to cover a journey of nearly three thousand miles."

[1] Burpee, Lawrence J., "Search for the western sea," p. 416.
[2] Burpee, Lawrence J., "Search for the western sea," p. 172.

· In 1792 Mackenzie ascended Peace river. He passed a post which had been built under his instructions in 1788 and up to which the river had already been surveyed on behalf of the North West Company, and continned upward to a point 6 miles above the mouth of Smoky river. After spending the winter at this point he set out in the spring of 1793 on that memorable voyage of discovery that was to carry him to the Pacific coast, ascending the Peace, the upper waters of which had never yet been explored.

A complete survey of Athabaska river appears to have been made by David Thompson, one of the greatest of Canadian explorers. In 1799 he descended Pembina river. On April 25 he reached the junction of the Pembina with the Athabaska, which he surveyed to the mouth of Lesser Slave Lake river. He then surveyed Lesser Slave Lake river to Lesser Slave lake, and returning to the Athabaska, travelled downstream to the mouth of the Clearwater.[1] In the spring of 1804 he surveyed the Athabaska from its mouth to the Clearwater and in 1810 explored a portion of the upper course of the river.

Among the early explorations that proved most fruitful in results were the two expeditions conducted by Captain (afterwards Sir) John Franklin. Accompanied by Dr. John Richardson, George Back, and Robert Hood, Franklin left Old Fort Providence on the north arm of Great Slave lake in 1820 on an overland trip to Coppermine river and the Arctic coast. He spent the winter near Winter lake erecting a small building which was named Fort Enterprise. In 1821 he descended the Coppermine and explored the Arctic coast eastward as far as Turnagain point. Returning he ascended Hood river as far as Wilberforce falls and from that point travelled overland to Fort Enterprise, reaching Fort Providence in December, after enduring indescribable hardship and suffering.

A few years later Franklin was placed in command of an expedition sent out to explore the Arctic coast west of Coppermine river. After spending the winter of 1825-6 on Great Bear lake the expedition descended the Mackenzie to the head of the delta. Here it divided, and Franklin, accompanied by Back, followed the western branch and explored the coast to Gwydyr bay near point Beechey. Richardson, with a young naval officer named Kendall, descended the eastern branch of the river and surveyed the coast eastward as far as Coppermine river. He ascended this river some distance and returned on September 1 to the fort on Great Bear lake where the winter of 1825-6 had been spent. Franklin reached the fort on September 21.

During the autumn of 1833 George Back explored a number of lakes lying to the northeast of Great Slave lake and in 1834 followed Great Fish river (now Backs river) to its mouth and explored a portion of the coast. Thomas Simpson and Peter Warren Dease, officers of the Hudson's Bay Company, carried on important explorations on the Arctic coast to the west of Mackenzie river in 1837 and to the east of the Coppermine in 1838 and 1839. In 1848 Dr. John Richardson again descended the Mackenzie and followed the coast to Coppermine river. Mention may be made also of the expedition of Anderson and Stewart of the Hudson's Bay Company down Backs river, and to that of Roderick MacFarlane in the Anderson River region.

[1] Tyrrell, J. B., David Thompson's narrative of his explorations in western America, p. LXXIX.

Explorations in the country to the north of Great Bear lake and between that lake and Great Slave lake were made by Emile Petitot, a missionary who was stationed in the Mackenzie basin for a number of years. More recent explorers are Warburton Pike and David T. Hanbury.

Important surveys and explorations have been made by members of the staff of the Department of the Interior, Ottawa, who made observations on the economic resources of the sections of country traversed. Our more intimate knowledge, however, of the geological features and mineral possibilities, has been obtained from investigations made by field officers of the Geological Survey, Canada.

Of the exploratory surveys carried out by the Department of the Interior those made by William Ogilvie are among the most important. During the summer of 1888 he entered Mackenzie basin from Yukon Territory by way of Porcupine river, Bell river, McDougall pass, and Rat river, a tributary of the Peel. He carried his survey up Mackenzie river, Great Slave lake, and Slave river to lake Athabaska, making observations on the natural features of the country through which he passed. He had already made surveys on the Athabaska and Peace rivers in 1884. In 1891 he ascended the Liard and surveyed Fort Nelson river, afterwards crossing to Fort St. John on Peace river. Explorations in the country to the east of Great Slave lake were made for the Department of the Interior in 1900 by J. W. Tyrrell; and the various surveyors who have run lines for the Department in the Athabaska and Peace River district have reported on the resources of the district.

Reconnaissance surveys have been made by the Geological Survey along the principal water routes.

In 1875 Alfred R. C. Selwyn explored Parsnip river and Peace river to the mouth of Smoky river and ascended Pine River South as far as Table mountain. John Macoun who accompanied him on this expedition continued the exploration down the Peace and up Athabaska river to the Clearwater returning by way of Methye portage, Fort Carlton, and Fort Garry.

In 1879 George M. Dawson made an exploration through northern British Columbia and Peace River district to Edmonton. Entering Peace River district by way of Pine River pass he descended Pine River South to its lower forks and from this point journeyed overland to Dunvegan. Going south from Dunvegan he explored part of Wapiti river, descended Smoky river to its mouth, and returned to Dunvegan. From this point he went overland in a southeast direction to Athabaska river, descended this river to Athabaska Landing, and ended his survey at Edmonton. R. G. McConnell, who was with Dawson on this expedition, travelled from Dunvegan by the north shore of Lesser Slave lake to Old Fort Assiniboine and to Edmonton. McConnell also descended Athabaska river from Athabaska Landing to La Biche river, and then made a traverse of the country by way of lac La Biche, Whitefish lake, and Goodfish lake to Victoria (now Pakan) on the Saskatchewan.

Robert Bell in 1882 made a geological examination of lac La Biche, La Biche river, Athabaska river below La Biche river, Clearwater river, and Methye portage.

The member of the Geological Survey staff who has carried on most extensive explorations in the Mackenzie basin and to whom we are most indebted for our knowledge of the geology is R. G. McConnell. In 1887 he descended Liard river to its mouth and at Simpson took passage on the Hudson's Bay Company's steamer for Fort Smith on Slave river. The remainder of the season was spent in examining the geology of Slave river, Salt river, Hay river, and of the western end of Great Slave lake. Winter was spent at Providence and from here rough traverses were made to lake Bistchô, Fort Rae, and other points. In the summer of 1888 he descended the Mackenzie to the Peel which he ascended to Fort McPherson. After making an exploratory trip up Rat river he crossed to Lapierre house by way of Peel River portage, descended Porcupine river, and carried on explorations in Yukon Territory.

In 1889 and 1890 McConnell explored the country between Athabaska and Peace rivers north of Lesser Slave lake. A careful examination was made of the geology along the Athabaska from Little Slave river to its mouth and along the Peace from Smoky river to Mikkwa river, and the more important tributaries were ascended.

The south shore of lake Athabaska was explored by D. B. Dowling in 1892, and the north shore by J. B. Tyrrell in 1892 and 1893. Cree river and the system of lakes and rivers to the east of lake Athabaska were also explored by Tyrrell and Dowling.

During the summer of 1899 explorations on Great Slave lake were conducted by Robert Bell and his assistant J. N. Bell. The latter remained at Resolution during the winter of 1899-1900 and in 1900 explored Great Bear river and the north and east shores of Great Bear lake. He made a traverse to Coppermine river, and returning to Great Bear lake reached Slave lake by a canoe route formed by a series of lakes and streams between McTavish bay and the north arm of Great Slave lake.

During the summer of 1902, Charles Camsell carried on explorations in the country to the west of Slave river, making an overland traverse southwest from Fort Smith to Salt river and Little Buffalo river, and following Jackfish river to the Peace. In 1905 he made a reconnaissance survey of Wind and Peel rivers. Entering Wind river by Nash creek he descended it to its confluence with the Peel which was descended to below Mount Goodenough.

In 1907-8, Joseph Keele made surveys on the Pelly, Ross, and Gravel rivers, descending the last named from its source to its junction with the Mackenzie.

Parts of the field seasons of 1913, 1914, and 1915 were spent by S. C. Ells in an investigation of the bituminous sands of Athabaska river. During the summer of 1914 Charles Camsell explored a section of country along Tazin and Taltson rivers between Athabaska and Great Slave lakes. In the same year a study was made by F. J. Alcock of the geology of a portion of the north shore of lake Athabaska. Further investigations were carried on by Camsell in 1915, by Camsell, Alcock, McLearn, Cameron, and MacVicar in 1916, and by Kindle, McLearn, and Cameron in 1917.

BIBLIOGRAPHY.

Following is a list of the most important publications that have been consulted in the compilation of this report.

Alcock, F. J.—"Beaverlodge Lake areas, northern Alberta and Saskatchewan," Geol. Surv., Can., Sum. Rept., 1916.

Back, George.—"Narrative of the Arctic land expedition to the mo
the Great Fish river and along the shores of the Arctic ocean
years 1833, 1834, and 1835. London, John Murray, 1836.

Bell, J. MacIntosh.—"Report on the topography and geology of Great
lake and of a chain of lakes and streams thence to Great Slave
Geol. Surv., Can., Ann. Rept., vol. XII, 1901, pt. C.

Bell, Robert.—"Mackenzie district," Geol. Surv., Can., Ann. Rep
XII, 1900, pp. 103A-110A.

British Columbia, Annual Reports of the Minister of Mines.

Burpee, Lawrence J.—"The search for the western sea, the story
exploration of northwestern America." Toronto, The Musson
Company, Ltd., 1908.

Cameron, A. E.—"Reconnaissance on Great Slave lake," Geol.
Can.; Sum. Rept., 1916.

Cameron, A. E.—"Explorations in the vicinity of Great Slave
Geol. Surv., Can., Sum. Rept., 1917.

Camsell, Charles.—"The region southwest of Fort Smith, Slave
N.W.T.", Geol. Surv., Can., Ann. Rept., vol. XV, 1903, pp. 151A-

Camsell, Charles.—"Report on the Peel river and tributaries, Yuko
Mackenzie", Geol. Surv., Can., Ann. Rept., vol. XVI, 1906, pt.

Camsell, Charles.—"The waterways of the Mackenzie River b'
Ottawa Naturalist, vol. 28, No. 2, May, 1914, pp. 21-33.

Camsell, Charles.—"An exploration of the Tazin and Taltson
Northwest Territories," Geol. Surv., Can., Mem. 84, 1916.

Camsell, Charles.—"Salt and gypsum deposits of the region be
Peace and Slave rivers, northern Alberta", Geol. Surv., Can.,
Rept., 1916.

Chambers, Ernest J.—"Canada's fertile northland, a glimpse o'
enormous resources of part of the unexplored regions of the Dom:
evidence heard before a select committee of the Senate of Ca
during the parliamentary session of 1906-7, and the report
thereon," 1908.

Chambers, Ernest J.—"The great Mackenzie basin," reports of the
committees of the Senate sessions 1887 and 1888, Dept. of the Int
Can., 1910.

Chambers, Ernest J.—"The unexploited west," Railway Lands Br
Dept. of the Interior, Can., 1914.

Clapp, Frederick G.—"Petroleum and natural gas resources of Can
Dept. of Mines, Can., Mines Branch, No. 291, 1915.

Crean, Frank J. P.—"New northwest exploration; report of explorat:
Saskatchewan and Alberta north of the surveyed area, seasons
and 1909", Dept. of the Interior, Can., 1911.

Dawson, G. M.—"Report on an exploration from Port Simpson o
Pacific coast to Edmonton on the Saskatchewan, embracing a po
of the northern part of British Columbia and the Peace River cou
1879", Geol. Surv., Can., Rept. of Prog., 1879-80, pt. B.

Dawson, George M.—"Notes to accompany a geological map of the nor
portion of the Dominion of Canada east of the Rocky mount:
Geol. Surv., Can., Ann. Rept., 1887, vol. II, pt. R.

Dawson, G. M.—"On some of the larger unexplored regions of Can
Ottawa Naturalist, vol. 6, May, 1890, pp. 29-40.

Department of the Interior, Can., Annual Reports.

DeSainville, Edouard.—" Voyage à l'embouchure de la rivière Mackenzie (1889-1894)", Bulletin de la Société de Géographie, vol. 19, 1898, pp. 291-307.

Ells, S. C.—"Preliminary report on the bituminous sands of northern Alberta," Dept. of Mines, Can., Mines Branch, No. 281, 1914.

Ells, S. C.—"Notes on clay deposits near McMurray, Alberta," Dept. of Mines, Can., Mines Branch, Bull. No. 10, 1915.

Ells, S. C.—"Investigation of bituminous sands in northern Alberta," Can. Min. Jour., vol. 37, No. 3, Feb. 1, 1916, pp. 73-74.

Ells, S. C.—" Bituminous sands of northern Alberta," Trans. Can. Min. Inst., vol. 20, pp. 447-459.

Franklin, John.—"Narrative of a journey to the shores of the polar sea in the years 1819, 20, 21, and 22, with an appendix on various subjects relating to science and natural history." London, John Murray, 1823.

Franklin, John.—"Narrative of a second expedition to the shores of the polar sea in the years 1825, 1826, and 1827, including an account of the progress of a detachment to the eastward by John Richardson." London, John Murray, 1828.

Footner. Hulbert.—"New rivers of the north, the yarn of two amateur explorers." New York, Outing Publishing Company, 1912.

Galloway, C. F. J.—"Report on the coal measures of the Peace River canyon," Ann. Rept. of the Minister of Mines, British Columbia, 1912, pp. 118-136.

Hanbury, David T.—"Sport and travel in the northland of Canada." London, Edward Arnold, 1904.

Harrison, Alfred H.—"In search of a polar continent, 1905-1907." London, Edward Arnold, 1908.

Hearne, Samuel.—"A journey from Prince of Wales's fort, in Hudson's bay, to the northern ocean, undertaken by order of the Hudson's Bay Company for the discovery of copper mines, a northwest passage, etc., in the years 1769, 1770, 1771, and 1772." Dublin, 1796.

Hooper, W. H.—"Ten months among the tents of the Tuski with incidents of an Arctic boat expedition in search of Sir John Franklin as far as the Mackenzie river and cape Bathurst.". London, John Murray, 1853.

Huntley, L. G.—" Oil, gas, and water content of Dakota sand in Canada and United States," Am. Inst. of Min. Eng., Bull. No. 102, June 1915, pp. 1333-1353.

Isbister, A. K.—"Some account of Peel river, North America," Jour. of the Roy. Geog. Soc., vol. 15, 1845, pp. 332-345.

Isbister, A. K.—"On the geology of the Hudson's Bay territory, and of portions of the Arctic and northwestern regions of America; with a coloured geological map," Quart. Jour. of the Geol. Soc. of London, vol. 11, 1855, pp. 497-520.

Keele, Joseph.—"A reconnaissance across the Mackenzie mountains on the Pelly, Ross, and Gravel rivers, Yukon and Northwest Territories," Geol. Surv., Can., No. 1097, 1910.

King, Richard.—" Narrative of a journey to the shores of the Arctic ocean in 1833, 1834, and 1835; under the command of Captain Back, R. N." In two volumes. London, Richard Bentley, 1836.

McConnell, R. G.—" Report on an exploration in the Yukon and Mackenzie basins, N.W.T.," Geol. Surv., Can., Ann. Rept., vol. IV, 1891, pt. D.

8

McConnell, R. G.—"Report on a portion of the district of Athabaska comprising the country between Peace river and Athabaska river north of Lesser Slave lake," Geol. Surv., Can., Ann. Rept., vol. X, 1893, pt. D.

MacFarlane, R.—"On an expedition down the Beghula or Anderson river," Can. Rec. of Sc., vol. 4 (1890-1891), pp. 28-53. Montreal.

Mackenzie, Alexander.—"Voyages from Montreal on the river St. Lawrenee through the continent of North America to the frozen and Pacific oceans in the years 1789 and 1793, with a preliminary account of the rise, progress, and present state of the fur trade of that country." London, 1801.

McLearn, F. H.—"Athabaska River section," Geol. Surv., Can., Sum. Rept., 1916.

McLearn, F. H.—"Peace River section, Alberta," Geol. Surv., Can., Sum. Rept., 1917.

McLeod, Malcolm.—"Peace river, a canoe voyage from Hudson bay to the Pacific by the late Sir George Simpson in 1828, journal of the late Chief Factor, Archibald McDonald (Hon. Hudson's Bay Company), who accompanied him." Ottawa, J. Durie & Son, 1872.

Macvicar, John.—"Foothill coal area north of the Grand Trunk Pacific railway, Alberta," Geol. Surv., Can., Sum. Rept., 1916.

Macoun, John.—"Geological and topographical notes on the lower Peace and Athabaska rivers," Geol. Surv., Can., Rept. of Prog., 1875-76, pp. 87-95.

Nair and MacFarlane.—"Through the Mackenzie basin, a narrative of the Athabaska and Peace River treaty expedition of 1899, by Charles Nair, also notes on the mammals and birds of northern Canada, by Roderick MacFarlane." Toronto, William Briggs, 1908.

Meek, F. B.—"Remarks on the geology of the valley of Mackenzie river with figures and descriptions of fossils from that region, in the museum of the Smithsonian Institution, chiefly collected by the late Robert Kennicott, Esq.," Trans., Chicago Acad. of Sc., vol. 1, 1869, pp. 61-114.

Milligan, G. B.—"Exploration survey in Peace River district," Rept. of Minister of Lands, British Columbia, 1914, pp. 90-95.

Moodie, J. D.—"Edmonton to Yukon, 1897," Report of the Northwest Mounted Police, 1898, pt. 2, pp. 3-82.

Ogilvie, William.—Report, 1884; Ann. Rept., Dept. of the Interior, Can., 1884, pt. 2, pp. 46-56.

Ogilvie, William.—"Exploratory survey of part of the Lewes, Taton-due, Porcupine, Bell, Trout, Peel, and Mackenzie rivers," Ann. Rept., Dept. of the Interior, Can., 1889, pt. 8.

Ogilvie, William.—"Report on the Peace river and tributaries in 1891," Ann. Rept., Dept. of the Interior, Can., 1892, pt. 7.

Petitot, Emile.—"Géographie de l'Athabaskaw-Mackenzie et des grands lacs du bassin arctique," Bull. de la Société de Géographie, sixième série, tome 10, pages 5-42, 126-183, 242-290. Paris, 1875.

Petitot, Emile.—"Autour du Grand Lac des Esclaves." Paris, 1891.

Petitot, Emile.—"Exploration de la region du Grand Lac des Ours. Paris, 1893.

Pike, Warburton.—"The barren ground of northern Canada." London, Macmillan and Co., 1892.

Preble, Edward A.—"North American fauna No. 27. A biological investigation of the Athabaska-Mackenzie region," Bur. of Biol. Surv., U. S. Dept. of Agriculture, 1908.

Reports of the Royal Northwest Mounted Police.

Richardson, John.—"Arctic searching expedition, a journal of a boat-voyage through Rupert's land and the Arctic sea, in search of the discovery ships under command of Sir John Franklin, with an appendix on the physical geography of North America." In two volumes. London, Longman, Brown, Green, and Longmans, 1851.

Roberson, William Fleet.—"Essington to Edmonton," Ann. Rept. of the Minister of Mines, British Columbia, 1906, pp. 101-131.

Selwyn, Alfred R. C.—"Report on exploration in British Columbia in 1875," Geol. Surv., Can., Rept. of Prog., 1875-76, pp. 28-86.

Report of the select committee of the Senate appointed to inquire into the resources of the Great Mackenzie basin, session 1888.

The Senate of Canada report, evidence and other documents presented by the select committee appointed to inquire and report from time to time as to the value of that portion of the Dominion lying north of the Saskatchewan watershed and east of the Rocky mountains, etc., 1907.

Russell, Frank.—"Explorations in the far north," University of Iowa, 1898.

Seton, Ernest Thompson.—"The Arctic prairies, a canoe-journey of 2,000 miles in search of the caribou; being the account of a voyage to the region north of Aylmer lake." New York, Charles Scribner's Sons, 1911·

Simpson, Thomas.—"Narrative of the discoveries on the north coast of America effected by the officers of the Hudson's Bay Company during 1836-39." London, Richard Bentley, 1843.

Tyrrell, J. Burr; assisted by Dowling, D. B.—"Report on the country between Athabaska lake and Churchill river with notes on two routes travelled between the Churchill and Saskatchewan rivers," Geol. Surv., Can., Ann. Rept., vol. VIII, 1896, pt. D.

Tyrrell, J. W.—"Exploratory survey between Great Slave lake and Hudson bay, districts of Mackenzie and Keewatin," Ann. Rept., Dept. of the Interior, Can., 1900-1901, pt. 3, pp. 98-131.

Vreeland, Frederick K.—"Notes on the sources of the Peace river, British Columbia," Bull. of the Am. Geog. Soc., vol. 46, pp. 1-24. New York, 1914.

GENERAL CHARACTER OF THE DISTRICT.

TOPOGRAPHY.

General Statement.

The Mackenzie basin, covering as it does some 680,000 square miles of territory in northwestern Canada, necessarily embraces a great variety of topographic forms. Its longer axis, about 1,350 miles in length, conforms in its direction to the trend of the main physiographic features and strikes northwestward. In width it ranges from 100 miles at the mouth of the river, to 900 miles near the centre of the basin.

It includes within these boundaries three main physiographic provinces (Figure 1) each of which runs almost the whole length of the basin and has characteristics which sharply distinguish it from the adjacent provinces. The three provinces are: a rugged, mountainous highland on the west, known as the Cordilleran region, which is a continuation of the mountainous region that forms the backbone of the North American continent; a rela-

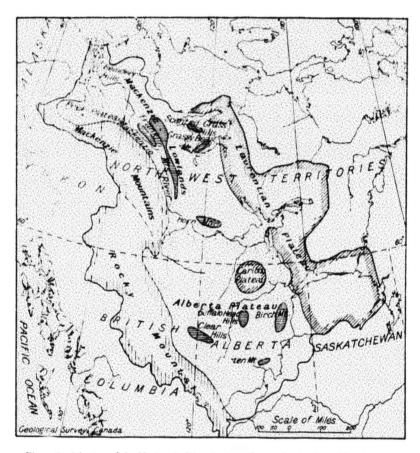

Figure 1. Diagram of the Mackenzie River basin, and its main physiographical features.

tively more subdued, but rocky, and partly treeless plateau in the east known as the Laurentian plateau and which is part of the great plateau that almost encircles Hudson bay; and between these two a broad, almost level, forested plain through which the trunk stream flows northwestward to the Arctic ocean, fed on the one hand by swift-flowing streams from the Cordilleran region and on the other hand from the numberless lakes of the plateau to the east. The last is the northward continuation of the Great Central plain of North America, and in this the Mackenzie river occupies the same position and performs the same functions that the Mississippi river does in the south as it flows towards the gulf of Mexico.

The drainage area of the Mackenzie river is a great basin, its western side dipping somewhat steeply northeastward and its eastern side sloping more gently westward to a central depression; and the whole tilted with a long easy slope northwestward to the Arctic ocean. The degree of slope of the central depression towards the northwest from the divide between the Saskatchewan river and the Mackenzie to the Arctic ocean is about 2 feet to the mile. This slope, however, is not uniform throughout, for besides the irregular hills and mountain ranges which rise out of the lowland there are north facing escarpments which interrupt and break the grades of the streams, and large local areas of depression in which the water accumulates to form the great lakes of this northern region. The largest of the lakes are Great Bear, Great Slave, and Athabaska, all of which rank among the large freshwater lakes of the world. The Laurentian Plateau region also holds innumerable lakes of smaller size on its uneven pitted surface.

Laurentian Plateau.

The great physiographic province known as the Laurentian plateau (Plate IIA), which covers such a great part of northeastern and northern Canada on both sides of Hudson bay, extends for some distance into the basin of Mackenzie river and occupies a strip along its eastern edge about 80 to 280 miles wide extending over a length of 800 miles from the height of land at the south to the northern shores of Great Bear lake.

The western border of this province, where it abuts against the central plain, is a fairly well-defined line marked by the contact between the Pre-Cambrian crystalline or metamorphic rocks and the flat-lying Palæozoic sedimentary rocks. This line of contact enters the Mackenzie basin from the south at Methye portage on Clearwater river in longitude 110 degrees west. Running northwesterly from there it passes the east end of Athabaska lake and follows the valley of Slave river to Great Slave lake. Crossing Great Slave lake in a northwesterly direction it runs from the northern end of the north arm of the lake to the southern point of McTavish bay on Great Bear lake and continues across the lake in the same direction until it passes outside the limits of the basin north of Smith bay. East of this line and extending far beyond the limits of the Mackenzie basin towards Hudson bay is the Laurentian Plateau province.

South of lake Athabaska there is apparently no topographic break to mark the change from the Laurentian plateau to the Great Central plain, the one merging gradually into the other. Between Athabaska and Great Slave lakes there is also no marked change of slope along the contact of the two provinces because of the thick deposits of alluvium that have been spread along that contact by the Peace and Slave rivers. North of Great Slave lake, however, the contact appears to be well defined by a low but abrupt escarpment, facing eastward and built of the flat-lying sedimentary rocks that underlie the Great Central plain. Whether or not this escarpment is continuous is impossible to say at present, but from the canoe route between Great Slave and Great Bear lakes it was noted by J. M. Bell[1] and Preble at a number of points. North of Great Bear lake enough in-

[1] Geol. Surv., vol. XII, p. 20 et seq.

formation is not available to define the nature of the contact between the two provinces.

The physical features of this province are typical of the whole Laurentian plateau generally. When viewed broadly the topography is that of a broad plain sloping gradually to the west and north with a gradient towards the great lake depressions which rarely exceeds 6 or 8 feet to the mile. Here and there residual round-topped hills or monadnocks rise a few hundred feet above the general level, but these hills are not as a rule connected into definite ranges nor aligned in any particular direction. In detail, however, the plateau is very irregular, broken, and rocky, with an uneven hummocky or mammillated surface (Plate IIA).

The surface of the plateau is nowhere broken by any prominent ranges of hills and the vertical relief is nowhere so great as that which obtains in the Cordilleran province. The greatest relief is found on the shores of the great lakes where it reaches a maximum of about 1,000 feet. On the north shore of Athabaska lake at Black Bay a group of irregular, round-topped hills rise 800 feet above the level of the lake[1], and on the northeast end of Great Slave lake, R. Bell[2] describes hills of granite and gneiss rising "as a sea of half-rounded hummocks to a general height of nearly 1,000 feet all along the northwest of this part of the lake and also around the northeastern extremity." Again, on the eastern shore of McTavish bay on Great Bear lake, steep cliffs rise abruptly from the lake shore to heights of 600 and 700 feet, and a few miles inland they reach a maximum of about 1,000 feet.[3]

Little is known of the character of the Laurentian Plateau province inland from the shores of the great lakes, except along a few canoe routes that have been traversed by Hearne, Franklin, Petitot, Tyrrell, Bell, Preble, Camsell, and a few others. From reports of these men it appears that the whole country is of a generally uniform level rising as a rule not more than 200 feet above the adjacent stream or lake levels and only at wide intervals is the evenness of the sky-line broken by any outstanding eminences. Instances of such eminences are, however, recorded by J. M. Bell[4] and Preble[5] in the country between Great Slave and Great Bear lakes where isolated peaks of conical or rounded outline rise as high as 1,000 feet above the level of the adjacent lakes. These higher hills are invariably composed of solid rock and usually of the harder igneous varieties; but lower hills of glacial origin and composed of sand, gravel, or boulders are described by J. B. Tyrrell,[6] in the region southeast of Athabaska lake. Such hills, however, rarely exceed 200 feet in height.

Throughout the whole 800 miles of Laurentian plateau lying within the basin of Mackenzie river the higher points show only slight differences in their altitudes above sea-level. For example, the maximum altitudes at Cree and Wollaston lakes in the southeast end of the basin are given by Tyrrell and Dowling as about 1,670 feet. On the north shore of Athabaska lake they are 1,490 feet, on Great Slave lake 1,520 feet, on the divide between Great Slave and Great Bear lakes about 1,700 feet, and on

[1] Camsell, Charles, Geol. Surv., Sum. Rept. 1914, p. 57.
[2] Bell, R., Geol. Surv., vol. XII, 1899, p. 106A.
[3] Bell, J. M., Geol. Surv., vol. XII, 1899, p. 18C.
[4] Geol. Surv., vol. XII, 1899, p. 23C.
[5] North American fauna, No. 27. U.S. Dept. of Agriculture, 1908, p. 114 et seq.
[6] Geol. Surv., vol. VIII, 1895, pt. D.

the east of Great Bear lake between 1,300 and 1,400 feet above sea-level. Taking the lower levels of the plateau, however, there is a much greater difference in altitude between the plateau at the southeast end and that at Great Bear lake. The difference here is about 1,200 feet, giving an average slope to the whole region of about 1·5 feet to the mile in a north-westerly direction.

The Laurentian Plateau portion of the Mackenzie basin is essentially a lake country, and its surface is covered with thousands of lakes of all sizes, ranging from mere ponds to lakes hundreds of square miles in extent. So numerous are these lakes and so rocky and irregular the country between them that the only method of travel used by the natives or travellers in summer in this region is by canoe. By portaging from one lake to the other it is possible to travel by canoe in almost any direction required, although the only routes in regular use are those leading from the trading posts to the hunting grounds or fishing resorts of the Indians.

The lakes are almost invariably shallow rock basins with smooth rocky shores and comparatively few beaches of sand, gravel, or boulders. They are usually very irregular in outline and their shape and alignment have been determined partly by the structure and composition of the rocks in which they lie and partly by the direction of movement of the glacial ice-sheet. Long continued decomposition of the bedrock of the plateau region and subsequent removal of the decomposed surface by continental glaciation are the chief causes in the production of a pitted, mammillated surface favourable to the accumulation of water as lakes.

Owing to the peculiar character of the surface of the plateau and the relatively short time that has elapsed since this character was developed, the rivers flowing through the plateau have no well-defined valleys nor have they evenly graded profiles. They are characterized rather by a succession of level stretches of river or lakes separated from each other by falls or strong rapids. Frequently they are merely a succession of lakes joined to each other by narrow, gorge-like openings at which falls or rapids occur which interrupt navigation. Their courses are very erratic and are frequently dependent on chance irregularities of the bedrock floor. In consequence, none of them in the plateau is navigable without interruption for any great distance, and in travelling on them portages have frequently to be made. They carry practically no material in suspension, for there is very little loose material on the surface of the plateau to transport and what they do pick up is quickly deposited again in the lakes through which they flow.

All the rivers of the plateau region drain to one or the other of the three great depressions, Athabaska, Great Slave, or Great Bear lakes, and owing to their peculiar characteristics, namely, lakes and waterfalls, they are excellent streams for the development of water powers.

The Laurentian plateau in the Mackenzie basin has as a rule little or no mantle of soil or other loose material covering its bedrock (Plate IIA), and there does not appear to be any large portion of it that would be available for agricultural purposes even if the climate were suitable. A very large proportion of the region has a rocky surface. Boulder clay is found frequently filling depressions on the surface and here and there occur sand-plains or other accumulations of glacial drift. The whole region has been subjected to intense glacial erosion by which the surface has been worn down to the live rock and denuded of its loose material

which has been carried westward and deposited in the lowland portion
of the basin. South of Athabaska lake is a large area underlain by horizontally bedded sandstone which on decomposition forms wide plains of
sand or gently rounded hills and ridges wooded with banksian pine.

In spite of its lack of soil, however, almost the whole region is wooded,
though sparsely, with a forest of spruce, banksian pine, tamarack, poplar,
birch, or willows. This forest becomes thinner towards the eastern and
northern edge of the basin and disappears entirely north of the east end
of Great Slave lake in the basins of Clinton-Colden, Aylmer, and Mackay
lakes. This is the only part of the Mackenzie basin included within the
great unforested region of northern Canada known as the Barren lands.

Cordilleran Region.

On the western border of the Mackenzie basin and extending throughout its whole length is a lofty mountainous region constituting a part
of the North American Cordillera. It forms a belt varying in width from
20 to 200 miles extending from the foothills which border the central
lowlands to the height of land separating the Mackenzie waters from those
which drain westward to the Pacific. The tributaries of the Mackenzie
cut deeply into the ranges which constitute this region and two of them,
the Liard and Peace river, cut right through, drawing much of their water
from the western or back slopes of the ranges and from the plateau
region lying to the west of them. This memoir, however, is not concerned
with that portion of the Mackenzie basin lying west of the summit of the
Rocky mountains which is drained by the upper waters of the Peace and
Liard rivers. The Cordilleran province of the Mackenzie basin as herein
defined, embraces the eastern slopes of the Rocky mountains, the Mackenzie
mountains, and Richardson mountains.

The eastern boundary of this province is a fairly definite line at which
the foothills of the mountains die out in the Great Central plain. Starting
in the south from a point near the intersection of latitude 53 degrees and
longitude 116° 30', the line runs northwestward, crossing the Peace river
near Hudson Hope and striking the Liard river near longitude 125 degrees.
Here, the continuity of the Rocky mountains is interrupted and they
appear to die away north of the river. Under the name Mackenzie mountains, however, the Cordillera springs up again north of the river, but
its eastern front is stepped far to the eastward and abuts against the
Liard river at Fort Liard. From this point the eastern boundary of the
Cordilleran province runs northward, touching the Mackenzie river at
the mouth of the Nahanni river and continuing thence along the western
side of the Mackenzie to latitude 65° 30', where it turns in a broad curve
and swings westward round the headwaters of Peel river. The Mackenzie
mountains die out about the head of Peel river in much the same way as
the Rockies north of the Liard river, but another lower range, known as
the Richardson mountains, springs up north of Peel river and extends
down to the Arctic coast, its eastern front following closely the valley
of Peel river and rising as an abrupt fault scarp out of the delta of the
Mackenzie.

This physiographic province of the Mackenzie basin comprises three
main mountain groups, namely, the Rocky mountains, Mackenzie mountains,
and the Richardson mountains.

Rocky Mountains. The Rocky mountains occupy the southwestern portion of the basin and extend as far north as the valley of the Liard river, where they either die out or lose their lofty mountainous character in a region of foothills. To the south they extend far beyond the limits of the Mackenzie basin into southern Alberta and Montana.

In the Mackenzie basin, as elsewhere, the Rocky mountains are made up of a series of parallel ranges striking northwesterly and coinciding in trend with the direction of the main mountain axis. Between the ranges are deep longitudinal valleys occupied by the smaller streams, and crossing them at right angles are transverse gaps through which the major streams break eastward to the Great Central plain. The transverse valleys are usually short in comparison with the longitudinal valleys and only in a few instances do such transverse breaks preserve their character so definitely through the whole breadth of the mountains as to form direct passes. The most notable instance of this is the Peace river, which cuts a deep, wide valley directly across the strike of the Rocky mountains and draws much of its water from the country to the west of them.

Some of the loftiest summits of the whole Rocky Mountain system in Canada are situated at the extreme southern portion of the Mackenzie basin at the headwaters of the Athabaska river, where streams which flow eastward to the Atlantic, northward to the Arctic, and westward to the Pacific interlock and have their sources. Here, is a group of mountain peaks which exceed 10,000 feet in elevation and culminate in the highest point of all, mount Robson, 13,700 feet above the sea. Between the Athabaska and Peace rivers not much is known of the character of the Rocky mountains except that they are very rugged and high and there are no known passes as low as the Yellowhead pass, 3,720 feet in elevation. Approaching the Peace river, however, the mountains decrease in height and width and are traversed by some low passes. Pine River pass, which was estimated by Dawson[1] to have an elevation of 2,850 feet above the sea, traverses the ranges where the bordering summits do not rise more than 6,000 feet above the sea; and Peace river itself cuts a valley directly through the main ranges of the Rockies, where their highest points barely exceed 6,000 feet in elevation and the valley bottom is less than 2,000 feet.

Less is known of the Rocky mountains north of Peace river than of any other portion of the whole system. Points as high as 7,500 feet are mentioned by McConnell[2] as occupying the region about latitude 57 degrees, but where the ranges have next been observed at the Liard river they are seen to have decreased in height to such an extent that the highest points are not more than 4,000 feet above sea-level. Only one of the main ranges of the Rockies is said to cross the Liard river, near Rivière des Vents, and even this a few miles north of the river becomes greatly reduced in height. The foothills, however, which border the main range on the east, do cross the river with undiminished strength at Hell Gate and probably merge with some of the southern spurs of the Mackenzie mountains to the northward.

The whole of this portion of the Cordilleran province is drained by three principal streams, the Athabaska, Peace, and Liard rivers. These rivers have their sources either in the central ranges of the Rockies or in

[1] Geol. Surv., Can., Rept. of Prog., 1879-80, p. 37B.
[2] Geol. Surv., Can., vol. VII, 1894, p. 16C.

the western slopes, but flow eastward by transverse ‹ ›s from the central ranges through a broad belt of foothills to the Great Central plain.

Mackenzie Mountains. Mackenzie mountains represent the Cordillera from the Liard river to the Peel and are the northwestern continuation of the Rocky mountains. They resemble the Rocky mountains in general characteristics and are made up of a series of parallel ranges striking north-westerly in the southern part and almost east and west in the northern part. "They are the greatest mountain group in Canada and appear to consist of two ranges, an older western range, against the eastern edges of which a newer range has been piled."[1] The newer range as well as a part of the older range lie on the Mackenzie River slope of the divide.

Mackenzie mountains have only been crossed in two places, namely, on Gravel river by Keele and on Wind river by Camsell. Their eastern front, however, was ascended and examined by McConnell at Liard river and North Nahanni river. At the south they rise somewhat abruptly out of a low-lying level region to heights of about 6,000 feet, and on Gravel river the highest summits reach a maximum of 8,000 feet. To the north they decrease again in elevation and on Wind river rarely reach 7,000 feet. They appear to die away at the headwaters of Peel river in a comparatively low region broken only by a few fault scarps and anticlinal ridges.

A number of important streams have their sources in and traverse Mackenzie mountains. Those tributary to the Mackenzie are the Nahanni, Root, Gravel, Carcajou, Arctic Red, and Peel rivers. These all cut across the strike of the ranges and their valleys are often continuous across the divide with those draining to the Yukon. All the streams have such high gradients that in no case do the natives make use of them when travelling from the valley of the Mackenzie into the mountains. They do, however, descend many of them in boats.

The divide at the head of Gravel river is given by Keele as 4,525 feet, but there are other passes leading to the headwaters of Macmillan and Stewart rivers which are said to be well below this level. The pass at the head of Wind river is estimated by Camsell at 3,400 feet. The vertical relief along the explored routes through these mountains ranges from 3,000 to 4,500 feet.

Richardson Mountains. North of Peel river, or from latitude 66 degrees, the Cordillera is represented by the Richardson mountains, a range which extends northward along the west side of Peel river to the delta of the Mackenzie where it swings westward as a coastal range bordering the Arctic coast.

These mountains are considerably lower in general elevation and have neither the length nor the breadth of Mackenzie mountains. They strike north and south and present other features which differ greatly from the other mountains of the Cordillera. McConnell[2] describes them as consisting essentially of two ranges separated by a wide longitudinal valley and flanked on either side by high plateaus. The eastern range has a width opposite Fort McPherson of 7 miles and its higher peaks are estimated to reach an altitude of 4,000 feet above the sea. They rise more or less easily from the valley of Peel river, but present a steep face about 3,000 feet in height to the delta of the Mackenzie. The western range is much

[1] Keele, J., "A reconnaissance across the Mackenzie mountains", Geol. Surv., Can., 1910, No. 1097.
[2] Geol. Surv., Can., vol. IV, 1888–89, p. 119D.

narrower and at Rat river does not exceed 4 miles in width, but spreads out to the south. The valley of Peel river, which skirts the eastern base, is fully 1,200 feet lower than Bell river on the western side and the drainage of the mountains is mostly towards the former.

The summits of this northern division are usually rounded in outline and show few sharp peaks, a feature which may be due either to mature erosion or to lack of close folding of the strata. The pass at the head of Rat river has an elevation of about 1,100 feet above the sea, and this is about the upper limit of tree growth in this part of the range.

The Great Central Plain.

This physiographic province occupies the central portion of the Mackenzie basin and includes all the region between the Laurentian plateau on the east and the Cordilleran region on the west. It extends throughout the whole length of the Mackenzie basin and has a length of about 1,300 miles. Its width ranges from a maximum of 420 miles along a line drawn across it through Fort Vermilion, to about 200 miles in latitude 63 degrees.

The Great Central plain of the Mackenzie basin is not as might be supposed a vast area of level country tilted with a uniform grade to the north, but its surface features are somewhat diversified. It has a general northward slope and along the valley of the main streams flowing through the plain the slope is generally uniform; but the Great Central plain is really made up of a northeasterly sloping plateau at the south, bounded by an escarpment or series of escarpments facing on to a lowland which extends northward to the Arctic.

South of the Mackenzie basin in southern Alberta and Saskatchewan, this plateau is referred to by Dowling[1] as the Cretaceous plateau since its limits are coterminous with the boundary of the Cretaceous rocks and its escarpment is built of Cretaceous strata. In the Mackenzie basin, however, the name is not appropriate since the escarpment of the plateau consists more often of Devonian rocks. It is proposed, therefore, to call this plateau of the Great Central plain the Alberta plateau, as its northern limits almost coincide with the boundary of that province. A smaller plateau occupies the upper part of the basin of Peel river and this has been called the Peel plateau[2]. The lowland portion of the Great Central plain is known as the Mackenzie lowland.[3]

Both the plateaus and the lowland have their surfaces broken by a number of hills or higher plateaus which rise from 1,000 to 3,000 feet above the level of the surrounding country.

The Alberta Plateau. This portion of the Great Central plain covers virtually the whole region south and southwest of Great Slave lake. It corresponds to the second and third prairie steppes in the Great Plains region south of the height of land[4] and though such a two-fold division might be made in the Mackenzie basin also, we have not as yet sufficient information of the topography to enable us to do so. The boundary of the plateau is marked by a series of north facing escarpments which extend from near Fort Smith on Slave river northwesterly along the south side

[1] Physical geography of Canada; Thirteenth report of the Geographic Board of Canada, 1914. See map.
[2] Camsell, Charles. Geol. Surv., Can., vol XVI, pt. CC, p. 23.
[3] Dowling, D.B., op. cit.
[4] Dowling, D.B., Physical geography of Canada: Thirteenth report of the Geographic Board of Canada, 1914.

of Great Slave lake and thence at an equal distance along Mackenzie river to the Liard in longitude 122 degrees. Here the escarpment is lost, although from McConnell's[1] observations it may swing southwest along the valley of Liard river. South of Fort Smith the plateau seems to merge with the Laurentian plateau.

Unlike the escarpment in Manitoba it is here not coincident with the eastern boundary of the Cretaceous rocks, but is built mainly out of Palæozoic limestones which overlie softer shales and so, in the course of erosion, give steepness to the slope. The smaller streams that break through from the plateau to the lowland all do so with high falls or strong rapids. Falls exist along this line on Little Buffalo river, Hay river, and Beaver river, and strong rapids occur at corresponding points on Buffalo river, Trout river, and the Liard. The larger streams, namely the Peace and the Athabaska, are, however, graded throughout and cut picturesque, terraced valleys in the plateau, that become gradually deeper and deeper to the southwest as the mountains are approached. For example, the valley of Peace river at Fort Vermilion is only about 100 feet deep, whereas at the mouth of Smoky river 300 miles farther upstream it is 700 feet deep and has a width of about 2 miles.

The top of the escarpment stands about 400 feet above the level of the adjacent lowland and the surface then rises gradually southwest and west to the foothills of the Cordillera. The slope, however, is so gradual that the smaller streams which have not the power to cut graded valleys from the plateau to the lowland, are comparatively sluggish in the plateau and are rapid and broken only where they descend through the escarpment. The surface, therefore, is monotonous and outcrops of the solid rocks are rare, and because the drainage is immature, muskegs are abundant and lakes fairly numerous. There are also many areas of open prairie land, especially in the basin of Peace river where settlement by agricultural communities is rapidly taking place.

The surface of the plateau is relieved by several higher plateaus or ridges which rise 1,000 or 2,000 feet above it. Of these, Caribou mountain, or as it should more correctly be called Caribou plateau, is situated between Peace river and Great Slave lake east of Hay river and covers an area of about 8,000 square miles. From the incomplete knowledge we have of its general character it appears to be a remnant of an older plateau, which is separated from the Laurentian plateau by the wide plains of Slave river above which it rises to a height of about 2,500 feet. The slopes on the north and east sides are said to be rather steep. On the south they are more gentle, and on the west they are still more so. The plateau is said to contain a number of lakes and is forested by a stunted growth of spruce similar to that which grows on the borders of the Barren lands or near the timber-line in the mountains.

· Buffalo Head hills which lie south of Peace river, rise abruptly from the plateau about 50 miles west of Wabiskaw river to an elevation of about 2,500 feet above the sea and, running in a south-southwesterly direction, die away opposite the mouth of Battle river. They are said by McConnell[2] to be about 50 miles long and from 25 to 35 miles wide.

Birch mountain extends for nearly 90 miles along the western side of Athabaska river below McMurray and has a width of about 35 miles.

---

[1] Geol. Surv., Can., vol. IV, 1885-89, pt. D.
[2] Report on the district of Athabaska, Geol. Surv., Can., vol. V, pt. I, p. 15D.

Its slopes on all sides rise easily to a maximum height of 2,300 feet above the sea, or about 1,000 feet above the surrounding country. The surface of the mountain is rolling and drift covered and the depressions are frequently occupied by lakes. It is all heavily wooded.[1]

Marten mountain, situated east of Lesser Slave lake and rising 1,000 feet above it, runs east and west for about 60 miles and is similar in general character to Birch mountain.[2]

Clear hills, north of Dunvegan, have a maximum height of 3,500 feet above the sea or 1,500 feet above the general level of the plateau, but their extent and general character are not clearly known.

Several other smaller ridges and plateaus, such as Pelican mountain and Thickwood hills, are described in the country south of Peace river[3] and there are no doubt many others which rise out of the plateau in the large tract of unexplored territory north and west of Hay river. All are erosion plateaus.

Peel Plateau. A plateau similar in character to that just described, occupies the upper part of the basin of Peel river.[4] This plateau extends south and west from the mouth of Satah river on either side of the valley of Peel river up to the base of the Richardson mountains, but its extent eastward is unknown. It is possible, however, that it forms a flanking platform to the Cordillera and extends east and south towards the valley of the Mackenzie.

On the north side, the Peel plateau forms an abrupt escarpment overlooking the lowland and rises gradually southward until it stands about 1,500 feet above the sea. Its surface is apparently flat but is in reality made up of several long undulations having a north and south trend. Peel river cuts a valley in the plateau, ranging from 700 to 1,000 feet deep.

Towards the upper part of Peel river the plateau forms a great bay in the mountains and in this bay its level is broken by several short ranges of mountains which are in reality outliers from the main Cordillera. These ranges are short, isolated hills seldom rising more than 2,000 feet above the plateau and frequently less than 1,000 feet. These hills enclose a shallow depression on the surface of the plateau which embraces the lower parts of the Wind and Bonnet Plume rivers. This depression was formerly a lake basin which, however, has since been drained, though the sediments that were deposited in it still remain to prove the former existence of the lake.

The eastern limit of the broken hilly country of the plateau lies at Snake river, beyond which the plateau stretches eastward with an unbroken sky-line for an unknown distance. It is everywhere covered with moss and wooded with small spruce and tamarack, and holds many small muskeg lakes on its surface.

The plateau gradually narrows in width northward as the Mackenzie approaches the mountains until it eventually merges with the mountains and disappears near Rat river at the head of the delta.

Mackenzie Lowlands. The lowland portion of the Great Central plain of the Mackenzie basin commences in the lower part of Peace river and the extreme western end of lake Athabaska, whence it extends as a narrow band down the valley of Slave river to Great Slave lake. Its eastern boundary

[1] McConnell, R.G., op. cit., p. 44D.
[2] McConnell, R.G., op. cit., p. 7D.
[3] Geol. Surv., Can., vol. V, pt. I, 1890-91, pt. D.
[4] Camsell, C., Geol. Surv., Can., "Peel river and tributaries", 1906.

here as well as north of Great Slave lake is the border of the Laurentian
plateau. It embraces the basin of the western end of Great Slave lake and
continues thence down the valley of Mackenzie river to the Arctic coast
its eastern boundary being as far as known the Laurentian plateau, and it
western Mackenzie mountains and the escarpment of the Peel plateau. ¿
re-entrant of the lowland extends up the valley of Liard river as far a
least as Fort Liard.

The limits of the lowland on the south are well defined and marked by ¿
distinct topographic break at the escarpment of the Alberta plateau
North of Liard river the lowland reaches on the west to the base of Mac
kenzie mountains which are here not bordered by an elevated plateau belt
North of Nahanni river, however, the long narrow ridge of Franklin moun
tains divides the lowland into two parts by separating a strip 20 to 80 mile
in width, through which the Mackenzie flows, from the main portion
of the lowland lying in the drainage basin of Great Bear lake. Franklin
mountains are believed to die out about latitude 66 degrees, and from thi.
line northward the lowland is bounded on the west by the escarpment o
the Peel plateau and farther north by Richardson mountains, whereas on the
east its limits are unknown though it probably extends across the height o
land into the watershed of Anderson river.

So far as it is possible to determine from our imperfect knowledge of the
eastern portions of the lowland there is no well-defined and continuous
break between the lowland and the Laurentian plateau such as occurs in the
west, but the plateau, as a rule, dips gently westward under the overlapping
edges of the flat-lying sedimentary strata which constitute the bedrock of the
lowland.

The elevation of the lowland at the west end of Athabaska lake is
about 700 feet above the sea and the slope of the surface from that point
to the Arctic averages about 5 inches to the mile and is so gradual that
from Fort Smith to the Arctic ocean, a distance of over 1,300 miles by the
river, steamers with a draft of 5 feet are able to run without encountering
any obstructions from rapids or falls.

The surface as a rule does not stand much above the main water
courses or the great lakes and the banks of these are rarely more than 200
feet in height.

The great bodies of water, Great Slave and Great Bear lakes, occupy
areas of depression in the surface of the lowland, and another long narrow
depression, not, however, now filled with water, lies along the base of
Mackenzie mountains from Fort Liard to the mouth of North Nahanni
river.[1]

Rising above the general level of the lowland are a number of hills and
mountain ranges varying in height from a few hundreds of feet up to about
4,000 feet. Such ranges are: Horn mountain northwest of Providence;
Franklin mountains, between the Mackenzie and Great Bear lake; Grizzly
Bear mountain and Scented Grass hills on the shores of Great Bear lake;
Reindeer hills east of the delta; and a number of other short ranges or
isolated hills.

These ranges are the only features that relieve the monotony of an
otherwise flat plain which holds on its forested surface innumerable lakes
and muskegs, and through which the smaller streams meander in shallow

---

[1] McConnell, R.G., Geol. Surv., Can., vol. IV, 1888–89, pp. 57D and 89D.

valleys. Only in the south along the valleys of Peace, Slave, and Liard rivers and the Mackenzie above the mouth of the Liard, is the land sufficiently well drained to be suitable for agricultural purposes. North of this, except on the islands and banks of streams, the surface is generally covered with a thick moss which prevents the ground beneath from ever thawing out beyond a depth varying from 6 inches to 6 feet. A thick forest of small spruce, tamarack, and willows extends over the whole of the lowland to the newly formed land at the seaward edge of the delta.

The southernmost of the ranges which rise out of the Mackenzie lowland is Horn mountain situated north of the Mackenzie between Providence and Simpson. The mountain is visible from the river all the way between these two points, at a distance of about 20 miles from the river. It has the character of a simple escarpment with a steep face to the south and a very gentle slope to the north.[1] Its summit stands about 1,000 feet above the sea.

The Franklin range, which extends along the east side of Mackenzie river from old Fort Wrigley northward beyond Great Bear river, is considered by McConnell to be an outlier from Mackenzie mountains which does not actually cross the river but marks an entirely new line of disturbance. The "Rock-by-the-river-side", however, near the present site of Wrigley, forms part of a small range which strikes in a direction diagonal to the main ranges and connects the Franklin range with the main portion of Mackenzie mountains.

The Franklin range at its south end rises easily out of the lowland as a narrow wooded ridge. It continues northward in a series of round topped eminences flanked in places by plateaus, and gradually increasing in height, culminates in the highest of them all, mount Clark. Mount Clark is a flat-topped mountain situated 8 or 10 miles back from Mackenzie river north of Blackwater river and is estimated by McConnell to be between 3,000 and 4,000 feet in height. From the northern shoulder of this mountain a line of lower elevations extends northward to Great Bear river and crossing it continues in a northerly direction for an unknown distance. A wide gap is cut in the range by Bear river at a point where the highest mountain, mount Charles, is 1,500 feet in height. Several other streams, the most important of which is Blackwater river, have their sources on the eastern side of the range and flow westward through gaps in it to join the master stream of the Mackenzie.

Another of the higher tracts that break the monotonous level of the lowland is Grizzly Bear mountain which occupies the peninsula between Keith and McVicar bays. The mountain is estimated by Sir John Richardson[2] to be 900 feet in height, but since many of Richardson's heights have been proved to be too low it is probably somewhat higher. Preble[3] considers it to be higher than mount Charles which Bell[4] gives as 1,500 feet in height. It is a massive, round-topped ridge quite devoid of trees for several hundred feet of its upper portion.

Scented Grass hills, on the north shore of the same lake, form the peninsula between Smith bay and the northern indentation of Keith bay known as Richardson bay. They are a round-topped ridge similar to

[1] McConnell, R.G., Geol. Surv., Can., vol. IV, 1888-89, p. 84D.
[2] Narrative, second expedition to polar sea, Appendix, p. ii, 1828.
[3] North American fauna No. 27, U.S. Dept. of Agric., 1908, p. 43.
[4] Geol. Surv., Can., vol. XII, 1899, p. 8C.

Grizzly Bear mountain in general character and height and terminating to the east in a prominent point known as Gros Cap.

Between the Franklin range and Mackenzie mountains, Mackenzie river flows through a forested plain 20 or 80 miles in width, which, however, is broken by several hills and plateaus of lower elevation than the bounding ranges. Several of these hills rise abruptly from the river to heights of 1,000 or 1,500 feet and because of their accessibility have been ascended and described by various travellers. The "Rock-by-the-river-side," also known as Roche Trempe-l'eau, 1,500 feet in height, has already been referred to as one of the points of a range connecting Mackenzie mountains with the Franklin range.

The isolated hill known as Bear Rock is another hill standing in the northern angle between Great Bear river and the Mackenzie. It is a flat-topped mountain rising 1,400 feet above the river and presenting its steepest face towards it. It runs in a northerly direction for about 4 miles and is the southernmost of a line of similar hills which extend northward along the valley of Mackenzie river. Another conspicuous elevation is Roche Carcajou situated about 100 miles below Bear Rock on the right bank of the river. It is about 1,000 feet in height and presents sheer cliffs rising for several hundred feet directly out of the water of the river. A weathered knob, whose summit from some points of view resembles the figure of a wolverine, gives its name to the mountain.

About 20 miles below Roche Carcajou the Mackenzie flows between two peaks of a low range, which are known as the East-and West-mountains-of-the-rapid. From this point, northward, the lowland through which the river flows is unbroken by any outstanding hills or ranges as far as the Lower Ramparts where the river cuts through the southern end of a low ridge. This ridge strikes north and rising in elevation in that direction becomes known as Reindeer mountains (Plate II B), or as referred to by Preble, Caribou mountains[1]. These mountains are described by Richardson[2] as following a course parallel to the east bank of the river. They rise to a height of 700 or 800 feet above the river, are uniform in outline, and have rounded summits. They diminish in height beyond latitude 68 degrees and gradually die out in the low land of the delta. To the northeast of their northern extremity a low escarpment rises abruptly from the east shore of Gull lake and faces west over the Mackenzie delta in the same way that the higher fault scarp of the Richardson mountains overlooks the delta from the west.

So much of the whole Mackenzie basin lying back from the main water courses yet remains to be explored that no doubt other topographic features than those just described will be discovered, which will tend to increase the diversity of a region at present considered to be of a generally uniform character.

RIVERS.

Athabaska River.

Athabaska river (Plate III A) drains an area of 58,900 square miles and has a length from headwaters to lake Athabaska of 765 miles. It rises on the eastern slope of the Rocky mountains near latitude 52° 30', its

[1] North American Fauna No. 27, U.S. Dept. of Agric., 1908, p. 34.
[2] Narrative Second Expedition to polar sea, Appendix, p. XXXVIII.

headwaters interlocking with those of Columbia and Fraser rivers about a group of high peaks of which mount Robson, 13,700 feet in height, is the highest. The passes through the mountains at the headwaters of the river are all high with the exception of Athabaska pass, 6,025 feet, and Yellowhead pass, through which the Grand Trunk Pacific railway runs at an elevation of 3,720 feet.

From its sources until it leaves Brûlé lake its course lies in the mountains where its grade is too steep for navigation. Beyond this point it pursues a northeasterly course, cutting a deep valley through the Alberta plateau and receiving in succession the waters of Baptiste river from the west, McLeod and Pembina rivers from the south, and Lesser Slave river from the northwest. The mouth of McLeod river, elevation 2,262 feet, is considered to be the upper limit of possible steamboat navigation, which then extends for a distance of 325 miles down to the head of Grand rapids. This portion of the river has an average fall of about 3 feet to the mile and is broken by a number of minor rapids, especially above the town of Athabaska, but none of them are sufficiently rough to obstruct the passage of steamboats during medium and high stages of water. Rapids also occur on Lesser Slave river, but in these the channel has been improved and deepened so as to allow steamboats to ascend the river to Lesser Slave lake. Lesser Slave river is one of the largest tributaries of Athabaska river and at ordinary high water stage has a discharge of about 2,300 cubic feet per second.

From Athabaska the river has a more northerly trend and although the stream runs with a strong current it is not broken by any rapids as far as Pelican rapids, a distance of 120 miles. Pelican rapids, however, present no difficulties to navigation except in low water when the channel is obstructed by some boulders. Broken water also occurs at two other points below Pelican rapids, but boats which are able to navigate Pelican rapids would have no difficulty in descending the additional 45 miles to Grand rapids which is the lower limit of steamboat navigation on this portion of the river. The river in this portion is from 250 to 400 yards wide and its valley varies from 300 to 400 feet in depth.

At Grand rapids the character of the river changes, its grade becomes greatly increased, and for the next 85 miles or as far as the mouth of Clearwater river it is broken by swift and dangerous rapids every few miles. The average grade of this portion is nearly 5 feet to the mile and the valley is 500 to 600 feet deep and frequently gorge-like. This portion is navigable only for scows or canoes and that with difficulty.

At Grand rapids (Plate III B) the river falls 50 or 60 feet in about half a mile and the main channel is not navigable by craft of any kind. A small island about a quarter of a mile long lies opposite the worst part of the rapids and over this goods are carried, and the scows are run down empty to the east of the island through a channel that has been blasted out among the boulders.

Rough water which requires some care in navigation extends for 2 or 3 miles below Grand rapids, but beyond this the river runs smoothly for over 20 miles to what are known as "Burnt" rapids. Here the stream widens to 400 yards and becomes shallow and occupied by scattered boulders. The canoe and boat channel follows the left side of the river and by keeping not more than 20 or 30 yards from the shore there should be no danger to experienced boatmen in the descent.

About 16 miles of smooth water follows Burnt rapids, below which river falls in quick succession over Boiler, Drowned, Middle, and Lo rapids, all of which occur within a stretch of 7 miles. They are all sim in character to Burnt rapids and owe their existence to a steepening grade and an accumulation of boulders in the channel. In Boiler ra the boat and canoe channels are on the left side and are so crooked t much care must be exercised to navigate them with safety. In Drowi rapids the channel is also on the left side and is somewhat rough. Abou mile below is Middle rapid which presents no great difficulties except th caused by shallow water and a few boulders. In this and Long rap which follows and is easily navigable, the channels are on the right side the river.

About 5 miles below Long rapid the river makes a sharp bend, at t end of which is Crooked rapid, where ledges of limestone project into t stream from the right side. The channel is on the right side where t only danger is from the large swells. Immediately below is Stony raj where the stream flows over a solid rock bed and is very shallow. T canoe channel follows the right shore.

At the foot of Stony rapid a crossing should be made to the left side order to descend Little and Big Cascades which follow. These are caus by ledges of limestone extending across the river, but they are broken dov in places enabling boats to get through. At Big Cascade a direct fall about 4 feet occurs on the right side, but on the other side this is resolve into a series of smaller drops and rapids that present no serious difficulty ordinary high water.

At Mountain rapid, 9 miles below, the river is again obstructed l ledges of limestone, but it is possible to avoid the rough water by descendin the upper portion on the left side, crossing in the middle over a piece c comparatively smooth water, and descending the remainder of the rapid o the right side.

Mountain rapid is the last dangerous rapid on the river, though a riff occurs 2 miles above McMurray at what is known as Moberly rapid.

Clearwater river, which joins the Athabaska from the east at McMurray is one of the largest tributaries of the Athabaska, but is navigable withou interruption for only about 40 miles, when strong rapids occur.

Below the mouth of Clearwater river, which has an elevation of 81 feet above the sea, the character of the Athabaska changes. Rapids di appear, and the Athabaska, enlarging to one-third of a mile or more i width, flows smoothly at an average rate of 3 miles an hour with a grade a far as Athabaska lake of about 8 inches to the mile. Islands becom numerous, the valley increases in width, and the banks gradually decreas from an elevation of about 400 feet at McMurray to the water level a Athabaska lake. The stream enters its delta about 35 miles from the lak and divides farther down into several branches whose channels al constantly changing as a result of the sediment and other materi carried down by the stream. The distance from McMurray to Atha baska lake is about 175 miles and the only difficulties to steamboa navigation in this distance are the sand bars and the shifting of the chann at the outlet.

Tributaries of Athabaska Lake.

With the exception of Peace and Athabaska rivers there are no streams flowing into Athabaska lake that are navigable for more than a few miles without interruption. On the south shore of the lake the most important streams are Old Fort, William or Gaudet, and Grand Rapids rivers. These rivers drain a large area of country south of the lake, but they are all shallow and so interrupted by rapids that they are navigable by canoe with difficulty, and that only in the high water stages.

On the nort! shore of the lake the streams are all short and none of them are navigable c .en by canoe, the reason being that the height of land between Athabaska lake and the streams flowing northward to Great Slave lake lies only a few miles north of the shore of the lake.

Stone or Black river which flows into the extreme eastern end of the lake is a large stream rising in Wollaston lake, at a height of 1,300 feet above the sea. Its course lies entirely within the Laurentian Plateau region and it has in consequence the characteristics of the streams of that region, that is to say numerous rapids or falls and lakes. On leaving Wollaston lake it flows northwestward through several small lakes almost as far as Water-found river in a country underlain by granite and gneiss. Here the stream enters an area of sandstone in which it cuts a shallow valley and is obstructed by numerous rapids and falls many of which have to be portaged. It enters Black lake at the east end by two channels and where it leaves it on the west side is 300 feet wide. Between Black lake and lake Athabaska the river falls 300 feet, mainly in three strong rapids at each of which a portage has to be made.

Peace River.

Peace river is formed by the junction of Finlay and Parsnip rivers and is the largest and longest of the tributaries of Mackenzie river. It rises on the western side of the Rocky mountains and, flowing eastwards, cuts through the axis of that range and drains a large area of country on its eastern slopes. The total area of its drainage basin is 117,100 square miles and its length from the sources of the Finlay to its junction with the Slave is 1,065 miles.[1]

The Finlay river rises near the headwaters of Skeena river and flows southeasterly down the great valley known as the Rocky Mountain trench, and the Parsnip, rising near the sources of some of the branches of the Fraser, flows in a northwesterly direction in the same great valley. On joining to form the Peace the latter turns at right angles and flows easterly in a deep valley which cuts directly across the main ranges of the Rockies.

From the confluence of Finlay and Parsnip rivers, which has an elevation of 2,000 feet above the sea, to Hudson Hope, a distance of about 120 miles, Peace river traverses the main ranges of the Rockies and their bordering foothills. Its valley averages about a mile in width and ranges from 2,000 to 4,500 feet in depth. The stream is from 300 to 500 yards in width and its current seldom exceeds 5 miles an hour. One rapid occurs just below the mouth of Finlay river and another, known as " Parle-pas " rapid, about 35 miles below.

[1] White, James, Atlas of Canada, Dept. of Interior, 1906.

For the last 25 miles of its course through the mountains and befo
reaching Hudson Hope the Peace flows through the canyon of the Mounta
of Rocks, to avoid which a portage 12 miles in length has to be made on t
north bank. The stream here contracts in places to about 150 feet
width and has a total drop of about 275 feet.

Hudson Hope is the head of the steamboat navigation for a stretch
about 550 miles of river, which extends from this point down to Vermili
chutes near the mouth of Mikkwa river. In this distance the stream has a
average fall of about one foot to the mile and the only obstructions
navigation are shallow bars which in low water make parts of the riv
unnavigable.

The width of the river varies from a quarter to half a mile and it
frequently divided by islands. The valley is cut to a depth of about 80
feet in the plateau, but this depth gradually decreases until at the Chut
it is not more than 100 feet. Its course as far as Peace River Crossing
westward and in this portion it receives a number of tributaries, namely tl
North and South Pine rivers and Smoky river. Smoky river is the mo
important tributary of the Peace, not only on account of its size at
length, but because it drains a large area of excellent agricultural territor
partly prairie and partly forested, which is rapidly being settled by ranche
and farmers.

From Peace River Crossing to Fort Vermilion the course of the riv
is northward, the valley is deep and picturesque, and the current had
uniform rate of 3 to 4 miles per hour. In the lower part of this stretch tl
valley widens, islands become more numerous, and the shores and ba
gradually change from gravel to sand or mud.

Wabiskaw river enters from the south about 20 miles below For
Vermilion and is the most important tributary below Smoky rive
McConnell[1] considers that this stream might possibly be navigated as fa
as Grand rapids, 9 miles above Panny creek, by powerful steamers with th
aid of a line, but with the exception of a few miles of still water above it
mouth it can hardly be considered a navigable stream.

At Vermilion chutes a rapid and fall with a total drop of 25 feet obstruc
the course of navigation and make a portage of 5 miles on the south ban
necessary. The chutes are caused by a ledge of limestone projecting acros
the river and though the rapid at the head is navigable, with difficulty, by
canoe, the falls below have a direct drop of 15 feet and cannot be run b
any sort of boat.

Mikkwa river, which joins the Peace immediately below the chutes, i
about 240 miles in length, but is a swift and shallow stream broken b
numerous rapids and only navigable for canoes during high water.

Below Vermilion chutes, Peace river increases in width and its valle
almost disappears. In the 220 miles to Slave river it falls about 110 fee
or 6 inches to the mile. Its current is consequently more gentle, and onl
at one point, a few miles below the mouth of Jackfish river, does it brea
into a rapid, which, however, is not a serious hindrance to navigation b
shallow draft steamboats. At this point and for 20 miles down, the strea
is confined by cliffs of gypsum from 20 to 80 feet in height, but at no othe
points are the banks composed of anything except sand or clay.

[1] Geol. Surv., Can., vol. V, 1890-91, p. 13 D.

Jackfish river is a stream 50 yards in width at its mouth, entering Peace river from the north, and is the most important tributary below Mikkwa river. It rises on the eastern slope of Caribou plateau and flows in a meandering course through a level country. It is navigable by canoes for about 200 miles with two or three interruptions.[1]

In ordinary stages of water Peace river discharges all its water directly into Slave river, but in flood time much of it flows by way of Quatre Fourches river into the western end of lake Athabaska and thence by Rocher river into the Slave. During these periods of flood the Peace becomes heavily laden with sediment. It has formed a great delta plain which has been gradually built out into the western end of Athabaska lake and so cut off a large body of water, lake Claire, from the main portion of the lake.

Stream measurements conducted at Fort Vermilion by officers of the Department of the Interior have shown that at that point Peace river has an extreme low water discharge in the month of February of about 7,500 cubic feet per second. In flood time in July the volume of water increases to about 300,000 cubic feet per second.

Slave River.

Slave river connects Peace river and Athabaska lake with Great Slave lake. A short portion of it, about 30 miles long, connecting Athabaska lake with Peace river, is known locally as Rocher river, but the name seems unnecessary. The total length of the river is about 300 miles.

Rocher river, the outlet of Athabaska lake, has an average width of about 150 yards and an easy current. A small rapid occurs on it about 10 miles above the mouth of the Peace, which, however, presents no difficulty to steamboats or canoes, except in extreme low water, when a portage frequently has to be made even with canoes.

Slave river after receiving the Peace becomes a broad and deep stream from one-third to two-thirds mile in width and has a current which runs at a uniform rate of about 3 miles an hour. To the head of the rapids at Fitzgerald or Smith Landing, a distance of 71 miles, the river flows between low wooded banks where solid rock outcrops only occasionally. No rapids of consequence occur in this stretch and no streams of importance enter.

From Fitzgerald to Fort Smith, 16 miles, the course of the river is interrupted by a series of dangerous rapids to avoid which a wagon road has been built on the west bank for the purpose of transporting the supplies for the Mackenzie River district. The rapids have a total fall of 125 feet, instrumentally measured, and are caused by a gneissic spur from the Laurentian Plateau region which crosses to the west of the river at this point. In these rapids the river spreads out to 1½ or 2 miles in width and contains many rocky islands. The descent of the rapids by boat is accomplished on the east side and five or six portages have to be made, but the course is so intricate and dangerous that it should not be attempted without the services of a good guide.

The rapids at Fort Smith constitute the head of navigation for steamers running in northern Mackenzie waters, which are unobstructed from this point northward into the Arctic ocean. If locks could be constructed here, permitting steamers to surmount the rapids, boats leaving the end of the

[1] Camsell, Charles, Geol. Surv., Can., vol. XV, 1902-03, p. 154 A et seq.

28

railway at McMurray could descend the river to the Arctic ocean, wit
run of 1,630 miles.

Slave river below the rapids is a monotonous stream winding throu
a level alluvial country, between banks of clay and sand which at F
Smith are about 100 feet high but become gradually lower northward. T
stream has an average width of about half a mile and a grade of abou
inches to the mile. The only important stream entering it is Salt river
stream about 30 yards wide at the month and draining a considerable tra
of country to the west, partly forest and partly prairie. In places, Sla
river meanders widely, forming bends which are several miles around l
only a few hundred yards across. For example, at Le Grand Detour t
river flows 15 miles around a big point, the base of which is about 1,0
yards across, whereas at point Ennuyeux the distance around is 10 mi
and the portage across is only half a mile.

The banks below Fort Smith are composed entirely of sand or clay a
only at two points do the solid rocks outcrop, namely, at Bell Rock,
massive cliff of limestone 7 miles below Fort Smith, and on the point belc
point Ennuyeux where beds of gypsum are exposed just at the wate
edge.

Slave river enters Great Slave lake by many channels through a del
of low alluvial islands, the delta having a spread across the outward si
of about 20 miles. An enormous amount of sediment carried by the riv
is deposited in the delta which is gradually being built out into the la
threatening to cut off the eastern and northern arms from the western ar

Tributaries of Great Slave Lake.

The principal streams flowing into the western arm of Great Slave lal
are Little Buffalo, Buffalo, and Hay rivers. These all enter from tl
south, and on the north no streams of importance are known.

Little Buffalo River.[1] This is a stream about 50 yards wide at i
mouth, entering Great Slave lake about 12 miles southwest of Resolutio:
It rises about 60 miles southwest of Fort Smith on the eastern slope of Cai
bou plateau and flows thence northeasterly in a shallow valley through a lev
forested country. Directly west of Fort Smith it turns northward and prec
pitates itself in a series of falls over a limestone escarpment which forms tl
boundary of the Alberta plateau, and for 7 miles below flows through
rocky gorge 100 feet in length.[2] Below this it parallels Slave river and a
Grand Detour approaches within a few miles of it. Here travellers by cano
frequently make a portage from Slave river and use Little Buffalo river as a
alternative and shorter route to Great Slave lake. The stream can l
navigated by canoe virtually to its head, but only in high water and l
making a portage at the falls.

Buffalo River. Buffalo river enters Great Slave lake about 60 mil
west of Resolution, and is stated by Ogilvie[3] to be about 100 yards wic
at its mouth. It is said to rise in the Caribou plateau, at the northern ba
of which it flows through a large lake, described and surveyed by A. l
Cameron[4] of the Geological Survey in 1917.

[1] McConnell, R.G., Geol. Surv., Can., vol. IV, 1888–89, p. 68 D.
Ogilvie, W., Rept. of Dept. of the Interior, Can., 1889–90, pt. VIII, p. 76.
[2] Camsell, Charles, Geol. Surv., Can., vol. XV, 1902–3, p. 159 A.
[3] Rept. of Dept. of the Interior, Can., 1889–90, pt. VIII, p. 76.
[4] Geol. Surv., Can., Sum. Rept., 1917.

From Buffalo lake to Great Slave lake is a distance of 75 miles, in which Buffalo river flows through a level country in a shallow valley 50 to 100 feet in depth. Rapids occur at several points, the most important of which are about 20 miles above Great Slave lake where the river descends a steep chute in a narrow limestone gorge, making a portage necessary when travelling by canoe.

Hay River. This river enters Great Slave lake 17 r. .i. s west of Buffalo river, and is the largest stream flowing into the west end of Great Slave lake. McConnell[1] describes it as rising near the headwaters of Fort Nelson river, the east branch of the Liard, and flowing in a northeasterly direction for 300 miles before entering Great Slave lake. The area of its basin is 25,700 square miles. Above Alexandra falls it is easily navigable by canoe and flows with a fairly uniform grade through a comparatively level, wooded country. Hay lake, an expansion through which the river flows, is said to be a large lake, though shallow, with low marshy shores. The river was surveyed by A. E. Cameron[2] in 1917, from the crossing of the 6th meridian down to Great Slave lake.

At Alexandra falls (Plate I) the river, which is 130 yards wide, precipitates itself over a limestone escarpment in two falls, the upper one of which is a sheer drop of 105 feet, whereas the lower one (Plate IVA), about a mile and a half below, has a drop of 46 feet in a series of steps. Below are 3 miles of rapids and a gorge 170 feet deep which continues for 5 miles below the lower fall. A portage of about 2 miles has to be made to avoid both falls.

The falls are about 20 miles in a direct line from the mouth of the river, but because of its windings the distance by the river is about 43 miles. The stream is swift but deep enough in high water for shallow draft steamers, and its valley is about a quarter of a mile wide. It forms a small delta at its mouth and enters Great Slave lake by two main channels.

The northern arm of the lake has only two streams of consequence flowing into it, namely, Yellowknife river on the east and Grandin river at the extreme north end.

Grandin River. At its mouth Grandin river is 50 yards wide and has a rather strong current. The first rapid is only a mile up and others follow in quick succession at which portages have to be made. The river has many branches, the main one being La Martre river which enters from the east and drains the large lake of the same name. Grandin river flows from the north through a rocky lake region and forms part of the regular Indian canoe route between Great Slave lake and Great Bear lake, a route followed and described by J. M. Bell[3] and E. A. Preble.[4]

Yellowknife River. This is a comparatively short stream, but is important as being the canoe route from Great Slave lake to the upper waters of Coppermine river. It lies entirely within the rocky Laurentian Plateau region and its course is consequently through a succession of lakes connected by rapids or falls at which portages are necessary. It was explored by Sir John Franklin in 1820, whose description of the river is still the best to be had.[5]

[1] Geol. Surv., Can., vol. IV, 1888-89, p. 69 D.
[2] Geol. Surv., Can., Sum. Rept., 1917.
[3] Geol. Surv., Can., vol. XII, 1899, pt. C.
[4] North American fauna, No. 27, U.S. Dept. of Agric., 1908, p. 108 et seq.
[5] Narrative of a journey to the shores of the polar sea in 1819-20-21 and 22.

The eastern arm of Great Slave lake has several streams emptyi
into it many of which are unknown beyond their mouths. The importa
streams are Taltson river on the south, Lockhart river at the east er
and Hoarfrost river on the north side.

Taltson River. Taltson river drains the greater part of the regi
between lake Athabaska and Great Slave lake. It rises in the Barr
lands near the headwaters of Thelon river and flows westward and the
northward through a series of lakes and connecting rivers until it empti
into Great Slave lake at the eastern edge of the delta of Slave river.
is not a navigable stream for anything but canoes and could only l
ascended by steamboats for 23 miles, to the first falls. Rapids and fal
occur at frequent intervals throughout its whole length and in travellir
on it portages have frequently to be made. Its principal tributary
Tazin river, a stream of similar character rising near the north shore
Athabaska lake. The total length of Taltson river must be about 4(
miles and its volume about 11,000 cubic feet per second. Tazin an
Taltson rivers are important as they form the canoe routes for the Indiar
of Athabaska and Great Slave lakes to and from their hunting groun(
on the edge of the Barren lands.[1]

Lockhart river rises in Mackay lake and flows successively throug
Aylmer, Clinton-Colden, Casba, and   illery lakes to join Great Slav
lake at its extreme eastern end. The   onnecting river between Macka
lake and Aylmer lake has not been explored, but is probably about 1
miles in length. From Aylmer lake to Casba lake the course lies entirel
within the limits of the Barren lands and there are no obstructions t
navigation. Between Casba lake, however, and Artillery lake a distanc
of 15 miles, the stream falls 32 feet and in going up stream three portage
have to be made, though only one is necessary going down. Artiller
lake is 55 miles long and from there to Great Slave lake, 24 miles, Lockhar
river falls 668 feet, so that it is nothing but a series of rapids and fall
that follow each other in quick succession. The highest falls, called by
Captain Back, Parry falls, have a drop of 85 feet into a rock-walled chasn
that varies from 20 to 50 feet in width. There are five other falls rangin
from 6 to 50 feet in height, which make the river quite unnavigable. I
consequence of this the regular canoe route follows a chain of eight smal
lakes lying to the east of Lockhart river.[2] The total distance from th
head of Mackay lake to the mouth of Lockhart river is about 300 miles

Hoarfrost River. The river empties into Great Slave lake on the norti
side by a fall 60 feet in height at a point where the adjacent hills are abou
1,000 feet high. The river has a very steep gradient and is full of rapid
and falls. It rises in Walmsley lake and forms a difficult canoe rout
through that lake to Artillery lake, a route that was followed by Captai
Back in 1833.[3]

Mackenzie River.

The name Mackenzie river is applied only to that portion of th
Mackenzie River system extending from Great Slave lake to the Arcti
ocean, a distance, according to Ogilvie, of over 1,000 miles. The tota

[1] Camsell, Charles, Geol. Surv., Can., Mem. 84, 1916.
[2] Tyrrell, J. W., Dept. of the Interior, Can., 1901, pt. III, p. 98.
[3] Back, Capt. G., "Narrative of Arctic land expedition in 1833-34 and 35"

distance, however, from the headwaters of the most distant tributary of
Mackenzie river, namely the Finlay, to the Arctic ocean, is 2,525 miles.

The Mackenzie on issuing from Great Slave lake has a width of 7 or
8 miles, but is shallow and filled with islands through which a moderate
current flows. Fifteen miles down, the islands cease and the river contracts to 4 miles. With a further decrease in width to 2 miles the strength
of the current increases to about 4 miles an hour, until at Providence, 45
miles down, several islands block the channel and cause an acceleration
of the current in what are called the " Providence rapids." In these
rapid the water, though swift, is quite smooth and steamers have no
difficulty in ascending them.

The country bordering the river below Great Slave lake is low and
flat and the valley is shallow, with banks seldom exceeding 30 feet in height.

Below Providence the Mackenzie passes through an expansion known
as the " Little lake " where it receives the water of a fairly large stream
from the north, Horn river, which rises in the rear of Horn mountain.
Continuing westward it remains wide and sluggish as far as a point near
Trout river known as the " Head of the line," so called because in ascending
the river the sluggish current above permits travellers to discard the
tracking line and use oars or paddles. Yellowknife river and Trout river,
streams that are both reported to head in large lakes, are passed on the
south side. The current increases in strength at the " Head of the line "
and from that point to the mouth of the Liard, 75 miles, it continues very
swift and the width of the stream is reduced to a little more than half a
mile. The banks are here slightly higher and instead of sand and clay
are composed of gravel and sand.

Simpson is situated on an island 2 miles long just below the junction
of the Liard and the Mackenzie. The main channel of the river is here
one mile wide and from this point northward to the Arctic the full width
of the stream is rarely less than this.

From Simpson to the mouth of Nahanni river distance of 75 miles,
the Mackenzie maintains its northwesterly dire.   'ts banks are about
200 feet high, with gravelly or bouldery beac   the current runs
at an average rate of 4 miles per hour. Several g   ...ps of long low islands
occupy this stretch and the only tributary large enough to be worthy
of a name is Martin river which enters from the southwest 8 miles below
Simpson.

At the mouth of Nahanni river the Mackenzie strikes against the
base of Mackenzie mountains and being deflected sharply to the north
by them, flows for several hundred miles parallel to and within sight
of their peaks. Several low islands, the longest 20 miles in length, occupy
the river bed below the Great Bend and behind one of these Root river
enters from the west. A few miles farther Willow Lake river comes in
from the east, and here the Mackenzie may be said to enter the mountains,
for a range, low and wooded at first, but gradually increasing in altitude,
rises out of the lowland on the east of the river and runs parallel to the
river for about 200 miles. This range is known as the Franklin range.

At the site of old Fort Wrigley, which is about 20 miles below Willow
Lake river, the Mackenzie is about 1½ miles wide, but below this point
for 100 miles high hills press closely down on either side and confine the
river to a channel free from islands and only about half a mile wide. The
current here increases in strength and runs about 5 miles an hour.

The present position of Wrigley is 25 miles below the old site almost opposite the " Rock-by-the-river-side " or " Roche Trempe-l'ea a steep, isolated, round-topped hill rising directly from the water's e to a height of 1,500 feet. Below this point the west bank becomes l and the east bank lower, and no feature of interest occurs for about miles, or until Blackwater river enters from the east. This strean about 75 yards wide and is easily recognized by the great volume of cle dark water that it discharges in to the Mackenzie and which preser its distinctive character for several miles before mingling with the wa of the larger stream.

Two miles below Blackwater river the Mackenzie turns shar to the west for 3 miles, below which it receives on the left a stream yards in width. Salt river is 33 miles down on the right side and Gra river 49 miles below Blackwater river. Gravel river is a mountain st flowing in from the west and is frequently used by the Indians as a r or canoe route from the mountains to the Mackenzie, but not in the reve direction.

The Mackenzie expands again about Gravel river, enclosing a num of islands, and from this point to the Ramparts, a distance of 326 mi the width of the stream is never less than a mile and frequently as m as 2 miles. The current varies in strength in different reaches but mu tains an average of about 4 miles an hour, and the banks are from to 400 feet in height.

A few miles above Norman, on the east side of the river, occasio columns of smoke indicate the presence of fires which are consum the seams of lignite outcropping in the bank. Landslides occasio by the burning out of the lignite seams and the undermining of the ban occur at several points, and at others the shales are baked and reddene These fires have been burning for at least one hundred and twenty-five yea or since Alexander Mackenzie made the first exploration of the river in 178

Norman occupies a commanding position on the east bank of the riv in the southern angle formed by the entrance of Great Bear river, a in the northern angle a steep, flat-topped mountain, known as Bear Roc rises to a height of 1,400 feet almost directly from the water of the Mackenz

"From the mouth of Great Bear river the Mackenzie runs in a gene west-northwesterly direction for 80 miles to Roche Carcajou, and th turns due west towards the East and West mountains of the rapid. this distance it has an average width of over a mile and occasionally expan around islands to over 10 miles in width. Its current is at the rate 3 or 4 miles an hour. Rugged limestone ranges are visible all along th reach on both sides of the river, but seldom approach within 30 mi of each other. The plains between and the lower slopes of the mountai are continuously clothed with forests of small spruce and aspen. T depression in which the river flows has a depth of from one hundred to fo hundred feet and a width of from 2 to 3 miles. River flats are seldo present and the banks of the valley slope more or less steeply up from t edge of the water."[1]

Roche Carcajou is about 1,000 feet high and rises steeply from t water's edge. A weathered knob, which from a certain point of vi resembles the figure of a wolverine, gives the name to the mountain.

[1] McConnell, R.G., Geol. Surv., Can., vol. IV, 1888-89, p. 102 D.

Twenty-five miles below Roche Carcajou the river falls over a ledge of rock to form the Sans Sault rapid, which is the most important obstruction to steamboat navigation from Great Slave lake to the sea. In high water the rapid is drowned out, but in low water the fall is increased and becomes a more serious obstruction, although not sufficient to prevent the passage of steamboats. The river is divided by a channel part way down the rapid and the usual canoe route is down the western channel where the fall is more gradual than on the east side.

Below Sans Sault rapid the river is much expanded and has an average width of nearly 2 miles, but on approaching the Ramparts it suddenly contracts to about 500 yards in width and bending to the east runs for about 7 miles between vertical cliffs of rock. At the upper end of the Ramparts these cliffs are 125 feet high, but increase towards the lower end to about 250 feet. The current in the Ramparts runs at a rate of 4 or 5 miles an hour and in low water a somewhat formidable rapid occurs near the upper end. The water throughout is deep and at the head of the Ramparts was found by Alexander Mackenzie to be 300 feet in depth.

Good Hope is situated on the right bank of the Mackenzie about 2 miles below the Ramparts and only a few miles south of the Arctic circle. The site has been occupied since 1836.

Hareskin river enters from the east 3 miles below Good Hope, and Loon river 21 miles farther. For the next 100 miles the Mackenzie maintains a northwesterly course and is never less than a mile in width and in what is known as the "Grand View" is about 3 miles wide. The valley is shallow and the river has many islands.

In about latitude 67° 30′ the river turns westward at right angles to the course it has previously been following and continues in this direction for about 60 miles, between high clay banks. The current here runs from 2 to 3 miles an hour. The country on either side is low and flat and no mountains are visible in any direction.

At the end of the western reach the river resumes its northwesterly course and its valley becomes deeper and more contracted for some miles, resembling a wide canyon. This has received the name of Lower Ramparts or the Narrows. The river here is half a mile wide at its narrowest point, but for most of the distance exceeds a mile in width. There is no sign of a rapid and the current is nowhere more than 5 miles an hour.

Arctic Red river enters from the west immediately below the Narrows and at its mouth is situated the trading post of the same name. Twenty miles below this Mackenzie river begins to spread out in its delta.

The delta of the Mackenzie extends in a north and south direction for about 100 miles and has a spread across its seaward side of about 70 miles. On the west the steep fault scarp of Richardson mountains rises abruptly out of the delta plain, and on the east it is bounded by a lower, rounded, range known as the Reindeer hills. Peel river joins Mackenzie river at the head of the delta and branches of both streams ramify in all directions through it, forming islands which usually contain many lakes. The banks are low, built of alluvial material, and forested with spruce and willows. Northward the banks gradually diminish in height and the forest growth decreases and finally disappears. Little is known of the depth of the water in the various channels of the delta, but Count de Sainville found a channel in the main eastern branch with not less than 5 feet anywhere throughout its length.

The Mackenzie is characterized by the comparative purity o
water, and since for its size it carries relatively little sediment its (
is not being built up as rapidly as that of its great counterpart, the M
sippi.

The great lakes within its drainage basin act not only as sediment:
basins, but also serve to regulate the flow of the stream. Its volu
therefore, does not vary very greatly at different seasons of the y
According to some rough measurements mad⁻ McConnell the disch
at a medium stage of water is about 500,000 c .u .eet per second.

The average grade of the stream for over 1,000 miles from G
Slave ..ake to the Arctic ocean is about 5 inches to the mile and at no p
does it greatly exceed this.

Liard River.

Liard river is one of the main tributaries of the Mackenzie. I
the Peace it has its sources west of the Rocky mountains and after unit
into one master stream cuts a deep canyon-like valley as it flows eastw
through those mountains. In its upper part it is divided into four nea
equal streams, namely the Kachika or Mud river, Dease river, Fra
river, and the branch which retains the common name, streams wh
interlock with the headwaters of Peace, Skeena, Stikine, and Yukon riv
It is a particularly rapid and dangerous stream having a fall betw
Dease lake and the Mackenzie of about 2,300 feet, giving an aver:
grade to the stream of over 3 feet to the mile. The descent of the ri
is greatest and its rapids most numerous while passing through and
some distance on either side of the Rocky mountains.

West of the Rocky mountains Liard river and its tributaries fl
through an irregular, forested plain, relieved at intervals by short disec
nected ranges of hills. Dease and Frances rivers, and the Liard abo
the mouth of the Dease, are all swift flowing streams, and navigable wi
difficulty even in ordinary stages of water. Below the mouth of t
Dease and as far as Rivière des Vents, which is at the western base of t
Rockies, the river is many times closely canyoned in and broken by rapi
The most formidable of these, namely Little Canyon, Cranberry rapi
Mountain Portage rapids, and Brûlé rapids, can only be overcome l
portages varying in length from one-half to 2 miles.

In its course through the Rocky mountains and the eastern foothi
the Liard is navigable only by canoe, and that with difficulty and co
siderable danger. In descending the river the first obstruction encounter
in this portion is Devil's rapids, where the river makes a wide be
to the northeast, all around which is a succession of rapids and canyo
At the lower end of the bend the river is reduced to a width of about
feet between rock walls. The portage trail across the bend is 4 mi
in length and runs over a ridge 1,000 feet in height.

"Below the Devil's portage for 30 or 40 miles the river flows throu
what is called the Grand canyon, but is more correctly a succession
short canyons, with expanded basins between filled with eddying curren
In low water the whole of this reach can be easily run in almost any ki
of a boat, but in the season of high floods . . . . the water forci
its way through the throat-like contractions is thrown into a commoti
too violent for any but the staunchest boats to stand. The canyon

reported to have been run in two hours, which would be at the rate of
about 18 miles an hour, an astonishing velocity, but the time was probably
underestimated."[1] The last of the constrictions in Grand canyon is
known as Hell Gate because it forms the entrance from below to the wild
portion of the river.

Shortly below Hell Gate the Liard breaks through the foothills and
enters the Great Central plain, in which it has an uninterrupted flow
and presents no obstacles to navigation until near its mouth. Its principal
tributaries on the north are Beaver and South Nahanni rivers, both streams
flowing from Mackenzie mountains and as yet little known. On the
south are Fort Nelson river, a comparatively easy flowing stream from
150 to 200 yards wide at its mouth, and Black river, a smaller stream
draining a lake country to the southeast.

Below Fort Liard the river flows in a shallow valley often filled with
islands and its current runs about 4 miles an hour. Within 30 miles of
the Mackenzie, however, the valley deepens and takes on the appearance
of a wide canyon. The current becomes swift and for nearly 10 miles
breaks over a succession of riffles. These are all easily run in a canoe
by keeping close to the right bank, but will form a rather serious obstacle
to navigation of the river by steamboats.

Before joining the Mackenzie the river expands to a width of nearly
2 miles, much of which, however, is occupied by islands and sand-bars.

Great Bear River.

Great Bear river[2] carries the water of Great Bear lake to the Mackenzie
in a stream about 90 miles in length. It issues from the lake in a broad,
shallow channel with a swift current which it maintains throughout its
course. It is confined between banks of sand or clay and is 150 to 200
yards in width.

About half-way down its course it cuts through Franklin mountains
and in doing so forms a rapid nearly 3 miles in length. This, however, is
easily navigable for canoes in all stages of water by following the right
hand shore, but by its shallowness presents considerable difficulties for
steamboats. Below the rapid the river expands and its current is not
so swift.

The water of Bear river is beautifully clear, of a greenish-blue colour,
and in marked contrast to that of the Mackenzie at its confluence with
the Bear.

Peel River.

The sources of Peel river and its upper tributaries lie on the northern
and western slopes of Mackenzie mountains. The main stream occupies
a great bay in the mountains and flows eastward as far as Snake river;
its principal tributaries join it from the south. At Snake river it swings
to the north and continues in that direction to its junction with the Mackenzie.

Peel river has not been explored above the mouth of Wind river, but
its course is said to be interrupted by falls about 30 miles above that point.

[1] McConnell, R. G., Geol. Surv., Can., vol. IV, 1888-89, p. 47D.
[2] Bell, J. M., Geol. Surv., Can., vol. XII, pt. C.

Wind river is a rapid, shallow stream with a discharge of about 5,00 cubic feet per second rising near the headwaters of Stewart river. forms part of a difficult canoe route from Yukon river to Mackenzie rive

Peel river is in a canyon at the mouth of the Wind, but a mile beloc expands to a width of nearly a mile. Seams of lignite are burning in th expanded portion, which extends for 15 miles down to the lower canyo Bonnet Plume river, which enters from the south 3 miles above the canyo is an unexplored river which at its mouth appears to be slightly larg than the Wind.

The lower canyon of the Peel is 2 miles long and about 500 feet dee with a current that runs from 6 to 8 miles an hour. It is formed by low range of hills crossing the river at this point. Below this the rive runs for 158 miles through the Peel plateau, cutting a valley which average about 800 feet in depth. The current is strong throughout, but no rapic occur.

Snake river is the largest tributary of the Peel, having a dischar considerably greater than the Wind river, and a width at the mouth about 350 feet. It is said to be navigable for canoes for a long distanc above the Peel.

At Satah river the Peel emerges from the plateau and enters th lowland of the Mackenzie, where the banks decrease in height to abou 100 feet and the strength of the current is reduced to about 2 miles a hour.

At Fort McPherson the Peel has a discharge in a medium stage o water of about 40,000 cubic feet per second and a velocity of about miles per hour. This is the highest point to which steamers ascend th river, but there seems no reason why stern wheel boats could not ascen the stream as far as the mouth of Wind river.

Shortly below Fort McPherson the Peel enters the delta of the Mac kenzie and although its eastern branch joins the Mackenzie 24 miles belov the Fort, the extreme western branch, locally known as Huskie rivei does not do so for about 100 miles more.

Other Tributaries of Mackenzie River.

Between Great Slave lake and the mouth of the Liard the Mackenzi receives a number of tributaries, but they are almost all unexplored beyon their mouths. On the southwest side are Beaver, Yellowknife, and Trou rivers, all of which are said to drain a lake country and to descend fror the Alberta plateau to the Mackenzie lowland in a fall or series of falls On the northeast side, the most important stream is Horn river whic rises behind Horn mountain and flows through a series of lakes to joi the Mackenzie at the "Little lake." It is about 50 yards wide at th mouth and is navigable for canoes for a long distance, but heavy rapid are stated to occur in its upper part. Other streams on this side ar Spence and Rabbitskin rivers, both as yet unexplored.

The important streams entering the Mackenzie between the Liar and Great Bear river are Nahanni, Root, Dahadinni, and Gravel rivei on the west side and Willow Lake and Blackwater rivers on the east, a of which, with the exception of the Gravel, are unexplored beyond thei mouths.

[1] Camsell, Charles, Geol. Surv., Can., "Peel river and tributaries." 1906.

Gravel river is described by J. Keele[1] as a swift mountain stream cutting across the ranges of Mackenzie mountains and entering Mackenzie river through a broad, alluvial flat. A rough measurement on July 19, 1908, above its mouth gave a width of 700 feet and a discharge of 25,000 cubic feet per second. It is used by the Indians as a boat route from the mountains to the Mackenzie.

Below Gre t Bear river the tributaries of importance are Carcajou, Hare Indian, and Arctic Red rivers.

Hareskin river heads near Great Bear lake and was partly explored by Macfarlane[2] in 1857. At its mouth it is about 200 yards wide and it retains that width for a long distance.

Arctic Red river is about 200 yards wide at its mouth and though shallow is said to be navigable for canoes for a considerable distance. It has, however, not yet been explored.

Tributaries of Great Bear Lake.

The country surrounding Great Bear lake is very imperfectly known and at present only two streams flowing into the lake, namely Dease and Camsell rivers, have been described.

Camsell river[3] enters an inlet on the south side of McTavish bay and drains a large part of the rocky Laurentian plateau to the south. It is important as forming part of the Indian canoe route from Great Bear lake to Great Slave lake. Its course is through a number of lakes and is broken by many rapids and falls so that it is only navigable for canoes by making numerous portages.

Dease river,[4] which discharges into the northeastern arm of the lake, is important only because it forms a difficult canoe route to Coppermine river. It rises in the treeless country between Dease bay and Coppermine river, and is a shallow, rapid stream not easily navigable even for canoes.

GREAT LAKES.

Athabaska Lake.

Athabaska lake, the southernmost of the great lakes of the Mackenzie basin, is a long and comparatively narrow sheet of water extending in an east-northeasterly direction. It has a length of 195 miles, a greatest width of about 35 miles, with a shore-line of over 520 miles, embracing a total area of about 2,900 square miles. It occupies a great depression excavated along the line of contact between horizontal Keeweenawan sandstones and the Pre-Cambrian schists and gneisses. Its surface stands at an elevation of 695 feet above the sea, but its depth has never been determined.

· The south shore of the lake is fairly direct and rises more or less abruptly to a great sandy plain, the maximum height of which is 500 feet above the lake level. The north shore is rocky, very much indented by bays, and friaged by hundreds of islands. The country to the north of the

[1] Geol. Surv., Can., "A reconnaissance across the Mackenzie mountains," 1910.
[2] Can. Rec. of Sc., Jan., 1890.
[3] Bell, J. M., Geol Surv., Can., vol. XII, pt. C.
[4] Hanbury, D.T., "Sport and travel in the northland of Canada."

lake is composed of irregular rocky hills which at Black bay rea
height of 800 feet above the lake. The divide between Athabaska
and Great Slave lake is only a few miles north of the former and
sequently there are few streams flowing into it from the north an
are short, rapid, and unnavigable. On the south side the streams, the
larger, are small, and none of them is navigable by canoes for any g
distance except Old Fort river. The principal inflowing streams
Stone or Black river at the east end, and Athabaska river at the
end. In periods of flood Peace river also flows into the lake by wa
Quatre Fourches river, but normally this and the Rocher river are a
flowing streams.

Athabaska lake formerly extended much farther west over the ba
of lake Mammawee and lake Claire, but Peace and Athabaska riv
have gradually built up deltas at their mouths and cut off these two la
from the main lake.

Towards the eastern end of the lake the water is clear and pure,
at its western end it is rendered turbid by the material discharged i
it by Athabaska river.

Athabaska lake is usually not free from ice until the middle of Ji
and it begins to freeze over again early in October. The western e
however, at the month of Athabaska river, is usually open early in May.

Great Slave Lake.

Great Slave lake, according to Cameron,[1] has a superficial ar
including islands, of over 12,000 square miles and ranks fourth am
the great lakes of the continent, being exceeded in size only by Superi
Huron, and Michigan. No comple. ey of the shores has yet be
made, but the lake is estimated to have a total length of 288 miles and
greatest width exceeds 60 miles.

Originally it had the form of a great cross, with one arm penetrati
eastward into the Laurentian plateau, two others stretched north and sou
along the contact between the Laurentian plateau and the horizon
sedimentary rocks of the lowland, and the fourth extended westwa
over the flat-lying rocks of the lowland. The southern arm extend
southward up the valley of Slave river, but has now been silted up by t
material carried in by that stream.

The eastern portion of the lake has a very irregular outline and
dotted by numerous rocky islands. Its shores are bold, rocky, a
sparsely wooded, and on the north side rise in places to a height of 1,0
feet above the lake. Into the north side, Hoarfrost river precipita
itself over a precipice 60 feet in height, and at the east end Lockhi
river enters, after flowing from Artillery lake in a series of rapids.
the south side the only tributary of importance is Taltson river which ent
near Slave river and drains most of the region between this arm of t
lake and Athabaska lake.

The northern arm of the lake is over 100 miles in length, but is narr
and filled with islands. Its shores are rocky and are nowhere very hi
Yellowknife and Grandin rivers are the only important streams dischargi
into it and neither is navigable except for canoes.

[1] Geol. Surv., Can., Sum. Rept., 1916.

The western arm presents a greater expanse of water, unbroken by islands, than either of the other divisions. "Its southern shore has a gentle, sinuous outline and is characterized by low banks and gently shelving beaches which are often thickly strewn with boulders. The banks . .. are often built up of drift timber. The northern shore is more uneven and is indented with several deep bays ". This arm is bordered on both sides by a flat, wooded country, which on the south side rises inland a few miles to some low rounded hills.

The water of Great Slave lake in the east and north arms is clear and probably deep. In the west arm, however, it is shallow and never entirely clear, as some of the sediment brought down by Slave river remains in suspension and drifts slowly westward towards the head of Mackenzie river.

Ice forms in the lake along the shores about the middle of October, but the middle of the lake frequently remains open until the end of November. In the spring, channels open along the shore early in June, but in the main body of the lake ice often remains until July 1 and in the east arm until somewhat later.

Great Bear Lake.

Great Bear lake is the next largest of the great lakes of the Mackenzie basin, having an area of about 11,400 square miles. Its shores have never been completely or accurately surveyed and for its delineation we are indebted mainly to Sir John Richardson[2] in 1825 and 1826 and to J. M. Bell[3] in 1900. It is very irregular in shape and is formed by the union of five arms in a common centre. Its greatest diameter from east to west is about 170 miles, and in a northeasterly direction from the bottom of Keith bay to the end of Dease bay is about the same distance. The greatest depth of water in the lake has never been ascertained, but in Smith bay a sounding by Bell[4] with 281 feet of cord did not reach bottom, and Richardson[5] states that 45 fathoms of line did not reach bottom in McTavish bay.

The shores of the la   in general low. Ranges of hills approaching 1,000 feet in height, h..   .er, occupy the peninsula between McVicar and Keith bays and that between Richardson bay and Smith bay. The shores of McTavish bay too are bold and rocky, often rising sheer out of the water for hundreds of feet. "The southern and western shores are well wooded, while its northern and eastern borders are more thinly forested. The immediate shores are mainly of sand and gravel and are usually devoid of trees, but are well clothed with willows and various ericaceous shrubs and herbaceous plants. In most places along the south shore this treeless stretch is only a few hundred yards in width, and in the bays the forest extends to the water's edge."[6]

The drainage basin of the lake is small in comparison with the size of the lake itself and there are consequently few streams flowing into it, all of which are short and rapid. Dease river at the bottom of Dease bay and

[1] McConnell, R. G., Geol. Surv., Can., vol. IV, 1888-89, p. 66D.
[2] Narrative second expedition to polar sea, Appendix, 1828.
[3] Geol. Surv., Can., vol. XII, pt. C, 1899.
[4] Geol. Surv., Can., vol. XIII, 1900, p. 99A.
[5] Op. cit., p. ii.
[6] Preble, E. A., North American Fauna, No. 27, U.S. Dept. of Agric., 1908, p. 43.

Camsell river at the southeast corner of McTavish bay are the best kn
tributaries.

The water of Great Bear lake is light bluish in colour and so clear
transparent that Richardson states a white rag dropped into the w
was visible until it reached a depth of 90 feet.[1]

Ice forms along the shores of the lake and in the bays early in Octo
but the centre of the lake is not frozen over until the beginning of Decem
The outlet of the lake, however, frequently rem ins open throughout
winter. By the middle of June the ice begins to break up along the sho
but drift ice sometimes obstructs navigation in the main body of the l
until the end of July.

Other Large Lakes.

Other lakes large enough to be of considerable importance tho
not large enough to be classed with the great lakes, are, lake Macl
Aylmer and Clinton-Colden lakes, and lac la Martre.

Lake Mackay is situated north of Great Slave lake on the edge
the Barren lands. It was discovered by Warburton Pike[2] in 1890
estimated by him to be about 100 miles in length, though he only
about 50 miles of it. The lake is cut by a number of long points and
numerous islands. The south shore is sparsely timbered, and the no
shore is entirely devoid of timber. The shores are all rocky and the h
at the northeast end are grouped in a regular range.

Aylmer and Clinton-Colden lakes[3] are only separated from ea
other by a narrow strait, known as the strait of the Sand hill. Th
combined length is said to be about 100 miles and the greatest width
miles. The shores are low and shelving, 80 to 100 feet high, and islar
are numerous. Both shores are practically devoid of timber, for t
lakes are both situated within the limits of the Barren lands.

Lac la Martre northwest of Great Slave lake is also one of the lar
lakes, but its general character and surroundings have never been lescril
by any writer.

Lesser Slave lake is 61 miles long and has a maximum width of
miles, embracing an area of 484 square miles. It is very shallow a
seldom exceeds 10 feet in depth at low water.

AGRICULTURE.

The division of the whole basin of Mackenzie river into three gr
physiographic provinces, namely, the Laurentian plateau on the ea
the Cordillera on the west, and between them the Great Central pla
makes it possible to delimit the possible agricultural portion of the ba
from that which is impossible of cultivation.

The Laurentian Plateau portion which forms about one-fifth of
whole basin and which is outlined on the accompanying physiograp
map (Figure 1), because of its rocky character, its absence of soil, a
its low average temperature, is not generally suitable for agricultu

[1] Narrative Second expedition to polar sea, Appendix, p. ii, 1828.
[2] Barren grounds of northern Canada, 1892, p. 58.
[3] Back, Captain, "Arctic land expedition, 1833-34-35," p. 138.
King, Richard, "Narrative of a journey to the Arctic ocean," p. 138.

purposes and can never be considered as part of Canada's reserves of agricultural lands. Except in a few localities where areas of clay land occur or where deposits of alluvium have accumulated no crops of any kind are at present being raised within the limits of this portion of the basin. Such localities are Chipewyan and Fort Smith where excellent crops of garden vegetables and some hardy grain are grown. These localities, however, are so close to the border of the Great Central plain as to partake very largely of the characteristics of that region.

In the Cordilleran portion of the basin only a very small proportion can be considered to have any agricultural possibilities, and this is confined to the main valleys such as the Peace and Athabaska and to the lower branches along these valleys.

The Great Central plain of the Mackenzie basin is pre-eminently that portion of the basin best adapted to agricultural pursuits (Plate VI). This portion of the basin constitutes about one-half of the whole basin, but it must not be assumed that the whole of it is occupied by cultivable land. So far as our incomplete knowledge of the country will allow us to draw any conclusions it may be said that the northern limit to which cultivation of the land may proceed in a large way is the line formed by Great Slave lake, Mackenzie river to the mouth of Liard river, and Liard river up to the mountains. This is approximately along the line of latitude 61 degrees north.

Northward of this line excellent crops of garden vegetables (Plate V) and barley have been raised along the valley of the Mackenzie at Wrigley and Norman, and even as far north as Good Hope on the edge of the Arctic circle. Similar crops might be grown in favourable localities in the country back from the river. In general, however, it may be said that the greater part of this portion of the basin inland from the streams is made up of spruce muskegs with little or no drainage and a permanent frost only a few inches beneath the surface, which would make farming impossible.

South of Great Slave lake and Liard river is an area within the Mackenzie basin covering about 200,000 square miles of the Great Central plain, which from the operations that have been carried on for years at the various trading posts of Fort Nelson, Fort Liard, Simpson, Providence, Resolution, Fort Smith, and Fort Vermilion, we know to be capable of supporting a population mainly by agriculture wherever the soil is suitable. Parts of this region are no doubt too high above sea-level, and other areas are low-lying spruce muskegs incapable of drainage, and cannot be considered as possible agricultural land. How great a part of this 200,000 square miles is to be considered unsuitable it is impossible to say, but it is no doubt large. The best part of it is believed to be that portion immediately adjoining the Peace river and its tributaries, especially Smoky river, and referred to generally as the Peace River country.

Within the limits of the Peace River country it has been estimated that there are areas totalling in all some 10,000,000 acres of prairie or slightly wooded country on which crops of vegetables and hardy grains can be successfully grown. The remainder of the country is more heavily wooded with poplar, spruce, or jack pine, and although the soil and climatic conditions are similar to those which obtain in the prairie portions it will necessarily be more difficult to clear and much longer before it is taken up by settlers.

Of the Peace River region the northern area which centres about F
Vermilion is apparently the most generally favourable for agricultu
Though more than two degrees farther north than Peace River Crossi
where so much settlement is now going on, the climatic conditions se
to be somewhat more favourable, due no doubt to the lower altitude,
country about Vermilion being about 1,000 feet above the sea and
plateau behind Peace River Crossing at least 1,000 feet higher still.

Fort Vermilion is the centre of a small farming community in wh
virtually all the vegetables, wheat and other grains, and meat necess,
for the support of the community are produced. Only a small proporti
of the arable land is, however, cultivated, because lack of transportati
facilities prevents the farmers from getting their produce to a market.

In the upper Peace River region an active immigration of settlers h
been going on for the last few years. Land is being taken up about
number of centres, mainly about Grand Prairie, Spirit River, Peace Riv
Crossing, and Dunvegan. Both farming and ranching are successful
carried on and since the entry of the Edmonton, Dunvegan, and Briti.
Columbia railway into that region in the summer of 1915 the settlers a
able to sell their produce in the markets of the world. In this region the
is said to be about 1,000,000 acres of actual prairie land and by the summ
of 1916 practically all of this had already been taken up in homestead
leaving only the partly or wholly wooded country for the future hom
steader.

In the basin of Athabaska river agricultural communities are growi
about Lesser Slave lake, Athabaska, lac la Biche, and McMurray, all
which except McMurray are already served by railways. Most of th
country is, however, a forested region, the percentage of prairie being muc
smaller than in the Peace River country, consequently development wil
be much slower. The crops grown in these localities are mainly vege
tables and the hardier grains, just as in the Edmonton district. Mixe
farming is generally practised throughout the whole district, but, as in th
Peace River country, only a small proportion of the total available land i
as yet taken up.

Farming and ranching operations on a small scale are being undertake
by the Roman Catholic Mission on Salt plain (Plate VI), a few miles wes
of Fort Smith. In this region there are several thousand acres of prairi
or partly open country covered with a variety of fodder grasses, makin
an excellent range for horses and cattle. It originally formed a part of th
range of the wood buffalo, but is now overrun by the horses of the resident
of Fitzgerald and Fort Smith, which winter out on the plain. The Missio
in 1916 had about 100 head of cattle and a few horses, and ground has bee
broken on which a variety of vegetables, barley, and oats were success
fully grown.

CLIMATE.

Since the basin of Mackenzie river extends over about 17 degrees o
latitude its climate is necessarily of such variety that it cannot be discusse
satisfactorily in a general way. The summer climate of the Mackenzi
basin, however, is not so much governed by latitude as the winter climate
for the isotherms for summer run in a northwesterly direction almos
parallel to the length of the basin, indicating a uniformity of temperatur

in that direction. The 55 degrees summer isotherm, for example, runs from Kenora on lake of the Woods northwesterly to the east of lake Winnipeg and thence through the east end of lake Athabaska, the middle of Great Slave lake, and the west end of Great Bear lake to Good Hope on the Arctic circle where it swings west into the mountains. The isotherms for the year, however, show a closer coincidence with the lines of latitude and indicate a decrease in temperature with an increase in latitude. This is due not so much to lower extremes of temperature in the northern latitudes as to a longer period of low temperatures in those regions throughout the year. The high latitude of, for example, Simpson at the mouth of Liard river, means a long cold winter and a short but warm summer. It involves also, however, a great increase in the amount of possible sunshine during the important growing period from May 15 to August 15. Compared with Ottawa, Simpson has an average of three hours more sunlight daily for the summer months, which means about eighteen days of additional sunshine during the three months when sunshine is most important.

In general it may be said that any point in the Mackenzie basin has a milder climate than any corresponding point of the same latitude in northern Manitoba, Ontario, or Quebec, probably because it is farther removed from the chilling influence of the large body of water, frozen for a great part of the year, contained in Hudson bay; also because it comes within the sphere of influence of the warm winds from the Pacific, and because of its soil covering.

Precipitation is fairly uniform throughout the whole basin and is nowhere excessive. It is in general slightly higher than that which obtains on the prairies of Alberta and Saskatchewan, but nowhere as heavy as it is on either the Atlantic or Pacific coasts. The total precipitation ranges between 15 and 20 inches annually and the snow in the central portion of the basin lies usually to a depth of about 2 feet. Snowfall is somewhat greater in the mountains to the west where the total precipitation too is higher.

The whole basin is wooded and although storms and blizzards occur occasionally they are by no means as violent as they are on the prairies because of the protection of the forests.

The great inland bodies of water, Athabaska, Great Slave, and Great Bear lakes, have a large influence in lowering summer temperature and retarding the advance of spring in the country adjacent to them because ice remains in them until late in June and in the northern lakes even until July. On the other hand, the influence of the warm winds from the Pacific in raising the temperature is felt along the eastern slopes of the mountain as far north as Liard river. These winds, commonly known as chinook winds, are particularly noticeable in the upper Peace River country and frequently bring warm waves during the winter months sufficient to melt the snow. They are not so prevalent in the summer months, but it was noticed by Dr. G. M. Dawson in the Peace River country in 1879, that summer frosts usually occurred on calm clear nights following a high wind from the west.

Temperatures during the day, even as far north as the delta of the Mackenzie, occasionally reach a maximum of about 85 degrees F. during the months of June, July, or August, but there is usually a decided drop as soon as the sun goes down. The nights, therefore, are always cool, with the temperature rarely above 50 degrees F., and except in the southern portion

of the basin it is liable to descend below the freezing point at any time of
year. During the winter the lowest temperatures occur in the month
January, when the thermometer occasionally drops in the southern par
the basin to 45 degrees below zero and in the northern part to 60 degr
below.

For the purpose of comparing the climate in different parts of the ba
the following tables of temperatures for the year, at a few widely separa
points, are submitted. Those for Athabaska, Peace River Crossing, F
du Lac, and Resolution are for the year 1912, and those for Simpson
Good Hope are for 1900.

Mean Daily Maximum.

Month.	Athabaska.	Peace River Crossing.	Fond du Lac.	Resolution.	Simpson.	Good H(
January.............	8·7	2·9	−16·2	....	−16·2	
February...........	28·3	24 2	− 3·4	....	− 9·2	
March..............	32·5	34·7	7·5	....	12·8	
April...............	51·7	52·5	32·9	30·3	41·2	
May................	65·6	70·0	50·3	53·7	52·6	
June...............	76·7	75·4	66·9	66·8	69·3	
July...............	68·5	71·6	66·5	63·6	69·4	
August............	69·6	72·1	....	66·3	65·0	
September..........	62·5	63·9	....	57·2	56·8	
October............	51·1	49·8	38·8	38·8	29·0	
November..........	36·6	32·7	25·5	21·0	5·4	
December..........	28·7	19·7	5·5	− 0·6	− 7·3	
Year............	48·4	47·5	....	....	30·7	

Mean Daily Minimum.

Month.	Athabaska.	Peace River Crossing.	Fond du Lac.	Resolution.	Simpson.	Good Ho(
January.............	− 9·7	−18·3	−33·2	....	−33·5	
February...........	3·6	− 2·1	−20·8	....	−33·5	
March..............	− 1·0	− 3·8	−18·1	....	−17·4	
April...............	27·7	27·0	11·0	11·7	17·8	
May................	34·0	35·7	29·7	30·0	35·2	
June...............	42·8	44·0	46·5	46·6	45·3	
July...............	43·0	44·5	47·0	45·7	45·7	
August............	46·8	45·4	....	49·6	44·1	
September..........	31·3	32·8	....	39·9	34·7	
October............	26·4	25·4	28·9	29·9	19·2	
November..........	17·9	13·4	12·7	12·8	− 8·7	
December..........	5·3	− 1·5	− 2·8	−14·1	−20·1	
Year............	22·3	20·2	....	....	10·7	

Extreme Maximum.

Month.	Athabaska.	Peace River Crossing.	Fond du Lac.	Resolution.	Simpson.	Good Hope.
January.............	39	33	12	....	5·0	11
February............	45	44	10	....	14·0	6
March...............	55	57	26	....	39·0	24
April................	65	64	51	54·0	60·0	48
May.................	87	89	67	77·0	66·0	56
June.................	93	93	90	88·0	80·0	80
July.................	83	88	79	77·0	79·5	83
August..............	81	88	....	79·5	78·0	75
September..........	81	82	....	78·0	71·0	60
October............	75	68	64	55·0	40·0	34
November..........	51	49	36	39·0	23·0	21
December..........	52	49	18	16·0	13·0	− 6
Year............	93	93	....	88·0	80·0	83

Extreme Minimum.

Month.	Athabaska.	Peace River Crossing.	Fond du Lac.	Resolution.	Simpson.	Good Hope.
January............	−47	−53	−51	....	−51·0	−62
February............	−34	−33	−34	....	−50·5	−46
March...............	−28	−30	−37	....	−50·0	−47
April................	21	17	−13	−13·0	5·0	−20
May.................	22	26	18	12·0	27·0	14
June.................	25	29	31	34·0	27·0	29
July.................	30	34	37	32·5	31·5	39
August.............	29	29	....	35·0	27·0	23
September..........	18	17	....	25·0	23·0	12
October............	8	8	15	13·0	3·0	−22
November..........	− 3	− 6	−11	− 8·0	−29·0	−47
December..........	−30	− 37	−20	−32·0	−48·0	−50
Year............	−47	−53	−51	....	−51·0	−62

Monthly Mean.

Month.	Athabaska.	Peace River Crossing.	Fond du Lac.	Resolution.	Simpson.	Good Hope.
January............	− 0·5	− 7·7	−24·7	....	−24·8	−35·7
February............	16·0	11·1	−12·1	....	−21·4	−26·1
March...............	15·8	15·5	− 5·3	....	− 2·3	− 7·6
April................	39·7	39·8	22·0	21·0	29·5	19·5
May.................	49·8	52·9	40·0	41·9	43·9	37·2
June.................	59·8	59·7	56·7	56·7	57·3	56·6
July.................	55·8	58·1	57·0	54·7	57·6	59·6
August.............	58·2	58·8	....	58·0	54·5	49·2
September..........	46·9	48·4	....	48·6	45·7	37·9
October............	38·8	37·6	33·9	34·4	24·1	12·3
November..........	27·3	23·1	19·1	16·9	− 1·7	−15·9
December..........	17·0	9·1	1·4	− 7·4	−13·7	−24·3
Year............	35·3	33·9	....	....	20·7	13·6

The freezing and breaking up of the rivers and lakes are events in mately connected with the climate and, therefore of interest in tl connexion. In all the rivers of the Mackenzie basin navigation is int rupted by drift ice in the autumn some time before the actual closi of the streams; and in the spring ice cakes continue to run for seve days after the breaking up of the rivers.

Athabaska and Peace rivers break up between the middle and t end of April, and ice may be expected to run in these streams any ti after the middle of October, though it may bé as late as the middle November.

Liard river is a week or ten days later in breaking up in the spri and a few days earlier when ice begins to drift in the autumn. The i from the Mackenzie below the mouth of the Liard is disrupted in spri through the influence of the Liard river, but it is usually from two three weeks before the break reaches the head of the delta of the Mackenz The Mackenzie above the mouth of the Liard generally remains solid f a week or ten days after the ice breaks up below that point.

Ice remains in Athabaska lake until about the middle of June, parts of Great Slave lake until the early part of July, and in Great Be lake until somewhat later. In the autumn, ice begins to form in the ba and sheltered parts of the great lakes early in October, but on accou of storms the main bodies of the lakes remain open for several weeks aft The small inland lakes and muskegs are frozen by the end of Octob in the southern par. of the basin and about a month earlier in the northe part. There is a difference, also, of about a month, between the southe and the northern parts of the basin, in the disappearance of the wint snow in the spring and the breaking up of the ice in the streams.

FAUNA.

Because of its wide range in latitude and the diversity of its top graphic features the basin of Mackenzie river contains a great variet of game and fur-bearing animals and birds, besides several species of foo fishes.

Moose (*Alces americanus*) are found throughout the whole regio northward to the mouth of the river and eastward to the limit of tree They are, however, more numerous in the central and mountain portior of the region than in the rocky Laurentian plateau to the east.

Barren Ground Caribou (*Rangifer arcticus*) are abundant at differer seasons of the year in different parts of the eastern border of the basi both in the barren grounds and the forested country bordering then They are in the main migratory in their habits and travel southward t the shelter of the woods in the autumn and north again in the early sprin

Woodland caribou (*R. caribou*) are not abundant anywhere and the range extends northward only to about latitude 62 degrees, but the mountai caribou (*R. caribou montanus*) are found in the Rockies and in the Mackenz mountains as far north as Gravel river.

Mountain sheep of three varieties (*Ovis canadensis, O. stonei,* an *O. dalli*) occur throughout the Cordilleran region, *Ovis dalli,* howeve being confined to Mackenzie mountains, *Ovis stonei* to the Rockies abou the head of the Liard river, and *Ovis canadensis* to the south of this.

Mountain goat (*Oreamnos montanus*) range throughout the Rocky mountains and are said to be found also in the southern part of Mackenzie mountains.

Muskox (*Oribos moschatus*) are confined to the barren grounds and are now even being restricted to the more inaccessible parts of that region. Within the Mackenzie watershed they occur now only about Aylmer and Clinton-Colden lakes.

Wood bison or buffalo (*Bison athabascæ*) formerly ranged over much of the country from Athabaska river to the Liard, but are now restricted to a few herds embracing some hundreds of individuals which inhabit the region west of Slave river (Plate VII) and north of the Peace.

The black bear (*Ursus americanus*) occurs more or less abundantly throughout the whole region and its range is coextensive with the forest.

Grizzly bears (*Urus horribilis*) are found throughout Rocky, Mackenzie, and Richardson mountains as far north as the delta of the Mackenzie.

The barren ground bear (*Ursus richardsoni*) frequents the barren grounds, but is not uncommon in the wooded country on the north and south shores of Great Bear lake.

The skins of polar bears (*Thalerctos maritimus*) are frequently obtained by the fur traders at the trading posts at the mouth of the Mackenzie, but the range of this animal is restricted to the coast and islands of the Arctic sea.

The basin of Mackenzie river is still the best fur producing part of the North American continent and a great many of the high grade furs are obtained from it. These include beaver (*Castor canadensis*), muskrat (*Fiber zibethicus*), lynx (*Lynx canadensis*) red, cross, black, and arctic foxes (*Vulpes*), otter (*Lutra canadensis*), mink (*Lutreola vison*), marten (*mustela americana*), fisher (*Mustela pennanti*), wolverine (*Gulo luscus*), wolf (*Canis occidentalis*), weasel (*Putorius*), and skunk (*Mephitis hudsonica*).

One of the most important animals in the region is the rabbit (*Lepus americanus*), for not only do the natives depend to a very large extent on it for food, but its skin also provides them with clothing and sleeping robes for winter use.

Most of the migratory game birds of economic importance in North America frequent the basin of Mackenzie river during the summer months. As soon as spring commences millions of ducks and geese of various kinds migrate northward to breed in the lakes and marhes of the region and of the coast and islands beyond. The marshes at the west end of Athabaska lake are a favourite halting place in this migration and large numbers of birds are to be found there in spring and autumn.

Spruce grouse, ruffed grouse, sharp tailed grouse, and ptarmigan remain in the region throughout the year and are found more or less abundantly throughout different parts of it.

The lakes and streams of the region abound in food fishes of various kinds. Whitefish, lake trout, pike, and sucker are found in nearly all the lakes and streams. Grayling, pickerel, and goldeye have a more limited range. Herring are found in the Mackenzie below the Liard and are abundant in Great Bear lake. The inconnu ascends Mackenzie and Slave rivers as far as the rapids at Fort Smith and is abundant in Great Slave lake the whole year round.

The only industry, apart from the agricultural one that is just be developed on the southern fringe of the basin, is that dependent on fur-bearing animals. Virtually the whole population, except the farm communities referred to, are more or less interested in the fur trade, a the few scattered settlements that are situated at intervals of 100 to ' miles along the main rivers were originally established, and are still ma tained, for the purpose of trading furs with the natives.

The fisheries, if we neglect a small industry on Lesser Slave la are still undeveloped except for purely local use. They are, howev one of the most valuable assets of the region. Thousands of whitefi are caught annually in Athabaska, Great Slave, and some of the smal lakes. In 1887, McConnell estimated a catch of 500,000 pounds of tl fish in Great Slave lake, to supply three posts, and from Athabaska la Ogilvie states that about 80,000 fish were taken in the autumn fishi to supply Fort Chipewyan alone. Great Bear lake, though containi the finest quality of fish of any of the northern lakes, is still untouche and besides Whitefish is known to be well stocked with lake trout and t Great Bear Lake herring.

TIMBER.

The whole basin of Mackenzie river is thickly wooded (Plate VIII with the exception of a narrow strip north of the east end of Great Sla lake, which forms part of the Barren grounds, and some areas of prair land in the upper Peace River region. The variety of trees, howeve is not great and includes only five coniferous and three deciduous specie namely, banksian pine, tamarack, white spruce, black spruce, balsar fir, and birch, aspen, and balsam poplar. All of these, excepting th banksian pine and balsam fir, extend practically to the limits of fores growth.

Banksian pine (*Pinus banksiana*) is the only species of pine in th region. It is found on sandy or gravelly ridges throughout the whol southern portion of the basin and extends down the valley of the Mac kenzie to latitude 64° 30', almost to the mouth of Great Bear river.

Tamarack (*Larix americana*) grows throughout the whole basi northward to the limit of trees and is found mainly in the muskegs. It wood is tougher than any other native tree and it is consequently used t some extent locally for purposes in which strength and hardness are re quired. It, however, does not grow in sufficient size and quantity t make it commercially important for lumbering.

White spruce (*Picea alba*) is the most important timber tree of th region and practically the only one that can be used for building or genera constructive work. Its range is northward to the limit of the forest an it even extends eastward along the valleys of certain rivers into the Barre grounds. It grows to best advantage along the banks and on the island of the streams and even as far north as latitude 61 degrees, on the Slave an Liard rivers, attains a diameter of 30 inches. Farther north along th valley of the Mackenzie it does not grow as large and there is a lesse quantity of merchantable timber, but even at the head of the delta o. Peel and Mackenzie rivers it reaches a diameter of 18 inches and a heigh of over 100 feet. Back from the river banks the tree is always smaller and except in a few sections of the southern part of the basin, there are few bodies of commercial spruce in the interstream areas. The rocky Laurentian plateau to the east has not as a general rule the soil conditions necessary to produce spruce timber of commercial dimensions. The best groves of spruce are certainly to be found in the valleys of the Great Central plain and of the Cordilleran region.

Black spruce (*Picea mariana*) has virtually the same range as the white spruce, but is confined mainly to the muskegs where it does not grow to large size. For this reason, and also because it makes a poorer quality of lumber, it is now of little use and not likely ever will be useful, except for pulpwood.

Balsam fir (*Abies balsamea*) is found along the valleys of Athabaska and Peace rivers and in the Cordillera for a considerable distance north, but it is not as abundant as the spruces.

Balsam poplar (*Populus balsamifera*) extends over almost the whole basin, but towards the mouth of the Mackenzie becomes very small.

The aspen (*Populus tremuloides*) also extends over the whole region and attains a large size along the valleys of the southwestern tributaries of the Mackenzie. Splendid groves occur in the valleys of Peace and Athabaska rivers, and it is one of the principal trees of the Peace River basin.

Birch (*Betula papyrifera*) is an important and widespread tree, the range of which is coextensive with the range of the aspen. It reaches its perfection in the southern part of the basin where trees 12 inches in diameter are common, but it gradually decreases in size northward. The tree is of great importance to the natives, who use its bark in the construction of canoes and baskets and its wood for toboggans, snowshoes, and handles of axes and other tools.

Considering the area of the Mackenzie basin and the fact that it is forested almost throughout, the quantity of merchantable timber is relatively small and is confined in all parts entirely to the stream valleys. The best areas are to be found on the tributaries of the Mackenzie which flow from the west, such as Athabaska, Peace, and Liard rivers. On Mackenzie river there is probably little more than is sufficient to supply the needs of the immediate vicinity and since it is a northward flowing stream its timber cannot readily be used for the settlements that are growing up in the southern part of the basin.

A great deal of valuable forest has been destroyed in p. t years, both through the carelessness of travellers with regard to camp fires and also through the deliberate starting of fires by the natives so as to improve the hunting. Forest fires are naturally more common along the regularly travelled routes, and in the unexplored portions of the country Indian canoe routes are often marked by the patches of burnt forest on the portage trails and at camping places.

Through the efforts of the Dominion Forestry department, however, the waste by fire is by no means as great now, and the natives are being educated to see the folly of allowing fires to spread. This waste is all the more deplorable since in that northern region tree growth is slow and reforestation consequently takes much longer than in more southern regions.

During the last one hundred and thirty years the principal route,
the transportation of supplies into the basin of Mackenzie river have char
a number of times, but in changing, the old routes have not been enti
abandoned and all of them are still used occasionally.

From the time of the first entry of the fur traders within the limit
the basin, in the last quarter of the eighteenth century, up to nearly f(
years ago, the main line of transportation was from the Churchill r
across the height of land at Methye portage to Clearwater river. Ha
entered the basin of Mackenzie river at that point the routes followed
main streams of Athabaska, Slave, Peace, Mackenzie, Liard, and l
rivers to the various trading posts situated on these streams, for all supp
were then carried in York boats propelled by crews of half breed voyage
or Indians. Winter travel also followed these routes.

When the Canadian Pacific railway was about completed, the route
the north country changed, and with Edmonton as a distributing poir
road was opened up across the height of land north of that point to At
baska river at Athabaska. Athabaska river then became the main
of travel, and in spite of the long series of obstructions in the rapids betw
Grand rapids and McMurray, all supplies were transported at consider-
risk and loss down this river, in scows, as far as McMurray. At t
point they were transhipped to river steamers and by these were car
and distributed to the trading posts, transhipments for the upper Pe
river taking place at Vermilion chutes and for the lower Mackenzie at Sm
rapids.

For the upper Peace River region west of Hudson Hope, the line
entry to the basin of Mackenzie river was from Fraser river across t
divide at Giscome portage and thence down McLeod and Parsnip rivers
the main Peace.

On account of the large volume of traffic in fur trading supplies tra
ported by the Athabaska River route, the first railway into the basin
Mackenzie was the Athabaska branch of the Canadian Northern railw;
which was built from Edmonton to Athabaska mainly to carry these suppli
This railway, with the Athabaska river north and northwest of it, continu
to be the main line of travel to the north until the year 1915, althou
many settlers for some years previous to that date travelled overland fr
Edmonton or Edson by way of trails or wagon roads that were being oper
up into the farming and ranching districts of the upper Peace River regi(

At present two other railway lines extend north or northwest into
basin of Mackenzie river. One of them, the Alberta and Great Wat
ways railway, is being built from Edmonton northward by way of lac
Biche to its proposed terminus at McMurray, but at the beginning of 1!
the end of steel was still about 90 miles from the terminus. When co
pleted this railway will be on the shortest and most direct route to the
northern interior of Canada.

The other line, the Edmonton, Dunvegan, and British Colum
railway, with its branch the Central Canada railway, runs northwestw;
into the agricultural districts of the upper Peace River country. 'l
northern terminus of the Central Canada branch is at present Peace Ri
Crossing, at which point connexion is made with the water transportat
systems which control the traffic beyond. The main line of the Edmont

Dunvegan, and British Columbia serves the agricultural country in the basin of Smoky river and will no doubt be extended farther west up the valley of Peace river.

Two transcontinental railways, the Grand Trunk Pacific and the Canadian Northern, run through the extreme southwestern part of the Mackenzie River basin between Edmonton and Yellowhead pass, crossing Pembina and McLeod rivers and paralleling a portion of the upper Athabaska.

During the season of navigation on Peace river, which lasts from May until the end of October, steamers make more or less regular trips from Peace River Crossing downstream to Vermilion chutes, a distance of 360 miles. Less frequent trips are made, mainly in medium and high stages of water, from Peace River Crossing upstream to Hudson Hope, a distance of about 200 miles.

At Vermilion chutes, where the Peace river has a total drop of 25 feet, steamboat navigation ends for this part of the river and a wagon road 5 miles in length has been built on the south side to transport supplies around the obstruction.

Below the chutes other river steamers make occasional trips, during May, June, or July, downstream to Athabaska lake or to Fitzgerald on Slave river. These steamers also ascend the Athabaska river to McMurray to attend to any traffic that might come either by the old Athabaska River route or the Alberta and Great Waterways railway. In connexion with these river steamers two small steamers run on Athabaska lake between Chipewyan and Fond du Lac in the interest of the fur traders who have posts at the east end of the lake, but like the steamers on the lower Peace and Athabaska rivers, no regular schedule of sailings is adhered to, as on the upper Peace.

The long series of rapids on Slave river between Fitzgerald and Fort Smith presents an obstacle to the Athabaska and Peace Rivers steamers which prevents them from continuing any farther north, and all goods have to be transhipped by wagon across a road 16 miles in length to steamers at the northern side of this obstruction.

From Fort Smith there is an unobstructed run of about 1,300 miles, to Fort McPherson at the head of the delta of the Mackenzie, by way of Slave river, Great Slave lake, and Mackenzie river, and recent soundings have shown that the steamers that make this run could continue out to the Arctic ocean nearly 200 miles farther through one of the eastern channels of the Mackenzie. The steamers of the Hudson's Bay Company and the Northern Trading Company make the trip to the delta of the Mackenzie once a season, but they are run on no regular schedule and solely for the convenience of these companies. The round trip is made in about four weeks. These boats each make a second trip later in the season either to Norman or Good Hope. This trip, made in August, is usually the last of the season, though navigation could be extended into September.

Accommodation is provided for passengers on all these boats, but if travellers propose getting into the country away from the main line of travel on the Mackenzie river they must make their own arrangements for transportation up the tributaries of the Mackenzie.

Besides those water routes over which steamers now run and which have a total length of river and lake shore-line of about 3,600 miles, there are, in the northern section of the basin, navigable waterways tributary to

Mackenzie river which have a total lengt.. of over **3,000** miles of
and lake shore-line.

The whole system of waterways is divided naturally into four sect
each of which is separated from the adjoining one by natural obstruc
of falls or heavy rapids which steamers cannot surmount. These sec
are: (1) Athabaska River section; (2) Peace River section; (3) Athal
Lake section; and (4) Mackenzie River section. These waterway.
tabulated below, the distances being given in round numbers becan.
lack of accurate surveys:

Navigable Waters of Mackenzie Basin.

Mackenzie River Section:

Mackenzie river, below Great Slave lake	1,200 miles.
Peel river, to mouth of Wind river	250 "
Bear river	90 "
Shore-line, Great Bear lake	1,360 "
Liard river	440 "
Shore-line, Great Slave lake	1,440 "
Slave river, Fort Smith to Great Slave lake	200 "
Total	4,980 "

Athabaska Lake section:

Slave river, Athabaska lake to Fitzgerald	100 miles.
Peace river, Slave river to Vermilion falls	220 "
Shore-line, Athabaska lake	560 "
Athabaska river, Athabaska lake to McMurray	170 "
Clearwater river	80 "
Total	1,130 "

Peace River section:

Peace river, Hudson Hope to Vermilion falls	550 miles.

Athabaska River section:

Athabaska river, Grand rapids to McLeod river	325 miles.
Lower Slave river and lake	115 "
Total	440 "
Total for whole Mackenzie basin	7,100 miles.

The Athabaska River section has a length of navigable river and l
of about 440 miles, on which steamers drawing 2 feet of water may run.

This section is separated from the Athabaska Lake section by
miles of rapids on Athabaska river, extending from Grand rapids
McMurray, which are navigable with difficulty for scows and canoes.

The Peace River section is 550 miles in length and extends fi
Hudson Hope down to Vermilion falls, and is navigable for steamers
a 2½-foot draft. The Wabiskaw river, a tributary of the Peace, in
section is said by McConnell to be navigable for powerful river stean
for a distance of 150 miles, but is not included in the table.

This section is interrupted at its upper end by the Peace River cany
where the river breaks through the Rocky mountains, and is separ
from the Athabaska Lake section by the rapids known as Vermilion chu
where there is a fall in the river of al t 25 feet. This obstruction cc
possibly be improved to such an exten. as to allow steamers to pass fi
the Peace River section into the Athabaska Lake section.

The Athabaska Lake section has 570 miles in length of navig
river, for boats of 2½-foot draft, and a shore-line on Athabaska lak

about 560 miles in length, making a total of 1,130 miles. It is separated from the Mackenzie section by a series of rapids on Sieve river about 16 miles in length, where there is a total fall of 125 feet. This break in navigation is now overcome by a wagon road of 16 miles from Fitzgerald to Fort Smith, but scows and light craft are usually taken down through the rapids by making four short portages.

The Mackenzie section is by far the most important of the whole system, covering as it does about 4,980 miles of known river and lake shore-line, on which a depth of water, ranging from 2 feet to 6 feet, may be found. This section embraces the trunk stream from Fort Smith down to the Arctic coast, a distance of 1,500 miles, over which a depth of 5 feet of water can be obtained. This, with the shore-line of Great Slave lake, 1,440 miles in length, and the small part of Peel river, is the only part of the section that is now being used by steamers. The remainder of the navigable waters of the section are only available for light draft steamers and cannot be navigated by the deep draft steamers that now ply on the portion previously mentioned. The Liard river is obstructed on its lower part by a strong rapid which, however, could be ascended by powerful light draft steamers with the aid of a line, making the navigable water on this stream 440 miles in length. Great Bear river, 90 miles in length, also has a shallow rapid about half-way up its course, which could possibly be ascended in the same way. With this obstruction removed or overcome, the whole of Great Bear lake, with a shore-line of about 1,360 miles, becomes connected with the Mackenzie system. Peel river is navigable for shallow draft steamers from the Mackenzie to the mouth of Wind river.

The Mackenzie has a number of other tributaries about which little or nothing is known, but which, on exploration, might prove to be navigable for certain distances. Among these are Little Buffalo river, Hay river, Willow river, Hareskin river, Arctic Red river, and some others.

COMMERCIAL POSSIBILITIES.

The natural resources of the basin of Mackenzie river are very varied, but as yet few commercial enterprises have been undertaken with a view to developing these resources. The oldest of these enterprises is that associated with the traffic in furs, but agriculture is now finding a large place in the commerce of the region. Transportation systems, both by railway and steamboat lines, have been developed to handle the products of both these industries; but, with the rapidly growing population and the large area of country capable of settlement that is still unoccupied, there is reason to believe that there will have to be soon a considerable increase in the transportation facilities of the region to meet its requirements.

In the southern part of the basin the agricultural industry offers probably the greatest possibilities, and though the value of the land for mixed farming, in the basins of Peace and Athabaska rivers, is becoming generally known, there is a large area of country north of this region, as far as latitude 61 degrees north, in which agriculture on a considerable scale could be carried on by hardy northern people. This northern portion is more difficult of settlement and the variety of crops that can be grown more limited, so that it will only be after the more southern lands are

tully taken up that the northern portion will be occupied by sett
There is, however, in this whole area of about 200,000 square miles, r
for a large population which ought to be able to support itself ma
or wholly by farming, ranching, or a mixture of both.

The fur trade has been carried on continuously within the Macke
basin since the latter part of the eighteenth century, and is still a large fa
in the industry of the region. Trading posts are scattered here and th
throughout the southern half of the I ......n among the agricult
communities, but in the northern half of the asin practically the en
population is actively interested in ' .... and exists by reason
its continuance. The history of the .... timately bound up v
the operations of the fur traders, and ....ered settlements situa
at intervals of 100 to 200 miles alon ......u streams were origina
established and are still maintained f ...... ose of trading furs w
the natives. Two principal companie-, ' ..... son's Bay Company a
the Northern Trading Company, div ...... rade in
northern half of the basin, and in the .... are a num
of smaller firms and individual trader. high grade fr
such as fox, beaver, marten, mink, lyn- ermi otter, are obtair
within the region and it is considered to be the ...... producing port
of Canada. Of the $5,000,000 worth of f s expe ....ually by Cana
the basin of Mackenzie river is said to supply nearly $2,000,000 wor
The fur-trading field, however, is now pretty well covered by the co
panies represented, though there are good opportunities for the individ
trapper to make a living.

In the mineral industry very little has been done except in the
western portions of the basin which lie outside the scope of this repo
namely in the placer gold fields of Cassiar and Omenica. Some prospecti
for oil and gas has been done along the valley of Athabaska river and al
recently on the Peace and strong flows of gas and some oil have be
obtained at several points. Prospecting for metallic minerals has al
been carried on to a very limited extent in the southeastern portion
the basin, and in the southwestern portion, good results have been obtain
in the case of coal and some coal fields located.

To appreciate the possibilities, however, from a mining point of vie
it is necessary to look at the geological map. Pre-Cambrian rocks occu
the eastern portion of the basin and deposits of iron, copper, nickel, and go
are recorded in them in certain localities. The mountainous weste
section is made up mainly of sedimentary rocks and with the excepti
of the extreme western part has not the possibilities in metallic minera
of the Pre-Cambrian area. Coal, however, occurs at a number of poin
and salt and gypsum at others, and placer gold is reported in some of t
streams In the Great Central plain the non-metallic minerals, co
salt, gypsum, oil, and gas, are known to occur, as well as the metal
minerals lead and zinc. Of these, the evidence seems to show that co
oil, and gas will probably be the most important, since the presence of a
at least is recorded in seepages, at intervals, from one end of the bas
to the other. Little or no attempt has been made to determine the val
of any of those mineral occurrences mentioned, and all that is known
the brief reference in the reports of explorers or travellers.

To the prospector, therefore, first of all, and later the mining ma
the basin of Mackenzie river affords a vast field of unknown but certain

great possibilities. Not only is the greater part of the basin quite unprospected, but more than one-third of it is still unexplored, and even its geographical features unknown. What these unprospected and unexplored regions contain in mineral resources it is impossible to say and unwise to conjecture about, in view of the developments that have already taken place in opening up other parts of northern Canada.

The food fishes of the streams and great lakes of the Mackenzie basin are another asset that will be utilized when methods of transportation are improved. At present the fisheries of lac la Biche and Lesser Slave lake are the only ones that have been developed to meet more than the local demands. Whitefish from these lakes are now shipped to points throughout western Canada because of the railway lines into these lakes. The more northern lakes, however, Athabaska, Great Slave, and Great Bear, contain an abundance of excellent whitefish, trout, and other food fishes, which are as yet not utilized except by the Indians and the few white people who reside on the shores of these lakes. A fishing industry of some magnitude will no doubt be developed on these lakes and at a number of other points, as soon as facilities for transportation are afforded where by the fish could be sold in the markets of the rest of Canada, for the cold waters of these northern lakes give a firmness and flavour to the fish which can not be excelled by the fish of more southern lakes.

Lumbering is not an industry in the Mackenzie basin that has or can assume such proportions as it has done either in eastern Canada or on the Pacific coast. The most important tree in the district is the spruce, which grows to a fairly good size in many of the river valleys, particularly those of Athabaska, Peace, and Liard rivers near the mountains. Though no doubt there will be a good local demand for lumber in the agricultural districts in the southwestern portion of the region, it will be difficult to take full advantage of the timber resources, partly because most of the streams flow northward away from the advancing settlement and partly because of the situation of the region in the interior of the continent. The commercial possibilities in lumbering, however, are to be considered.

Water-power will no doubt be important in the commercial development of the region because of the scarcity of coal. As in the Laurentian Plateau regions elsewhere, the rivers of the eastern part of the Mackenzie basin contain a number of falls at which power could be developed. Among those which have been noted by explorers are Black river at the east end of Athabaska lake, and Taltson, Lockhart, and Yellowknife rivers flowing into Great Slave lake.

Water-power could also be developed on Peace river at the chutes, where there is a total fall of 25 feet, part of it being a direct drop of 15 feet, and on Slave river near Fort Smith where the river falls 125 feet in a series of strong rapids spread over a distance of about 14 miles. The most evident water-powers and those which no doubt could be most cheaply developed occur on a number of rivers which fall over the face of the Devonian escarpment south of Great Slave lake. Little Buffalo river, for example, has a total fall over the escarpment of about 100 feet, half of which is a direct drop. Hay river also falls over the escarpment in two leaps with a total drop of 151 feet, the upper fall having a sheer drop of 105 feet and the lower one 46 feet. Beaver river, which flows into the Mackenzie immediately below the outlet of Great Slave lake, plunges in a direct fall over the same escarpment, and the two next streams north,

Yellowknife river and Trout river, are also said to fall in the same over the same escarpment. Water-power could be developed on a nu of other streams, but a sufficient number have been mentioned to s that there is much available water-power to meet the requirement. any mining, lumbering, or other industries that might spring up in region.

The white people of the Mackenzie basin are few in number : north of Peace river and Athabaska lake consist chiefly of the fami of the fur traders and the missionaries of the Protestant and Ron Catholic churches. These reside at the small posts (Plate IXA and which are scattered at intervals of 100 to 200 miles along the valleys Slave, Liard, and Mackenzie rivers, and their total number is very sm A considerable number of white settlers, however, have within the l five years entered the southwestern part of the Mackenzie basin and ta up homesteads in the upper part of Athabaska River basin, the basin Smoky river, and other neighbouring parts of Peace river. The build of several lines of railway into these regions has been the cause of t large influx of people.

The natives of the region include Esquimo, and Indians of th different stocks, Athapascan, Cree, and a few Iroquois.

The Esquimo are confined to a belt along the Arctic coast, but numbers that live or trade at posts within the Mackenzie basin can be much more than 150.

The Iroquois consist of a few families living about the headwat of Smoky and Athabaska rivers, descendants of boatmen and voyageu of the fur traders.

Crees occupy the greater part of the basin of Athabaska river a1 the region between that stream and the Peace.

The remainder of the Mackenzie basin is occupied by various trib of the Athapascan family, comprising in that region about 6,000 to 7,0(individuals. These tribes are situated as follows: Chipewyans in the regi south and east of Athabaska lake, Caribou-eaters north of Athabasl lake, Yellowknives east of Great Slave lake, Dogribs north of Great Sla· lake, Hares north of Great Bear lake, Beavers in the upper part of Pea river and Fort Nelson river, Slavis in the valley of the Mackenzie fro Great Slave lake to Great Bear rive·, Loucheux in the basin of Peel riv(and Sikannis, Kaskas, Nahannis, and other mountain tribes in the Rocl and Mackenzie mountains. These and the other native tribes live I hunting and trapping and very much in the same manner as they ·d before the advent of the white people. They have, however, all becor christianized and are adherents either of the Church of England or tl Roman Catholic church.

GENERAL GEOLOGY.

The Mackenzie basin is underlain in part by rocks of Pre-Cambri; age, forming the western edge of the great Canadian Shield, but to a mu(greater extent by sediments of later deposition. The sediments ha·

suffered very little deformation and, except in the mountains on the western edge of the basin, display only a few minor elevations.

The Pre-Cambrian rocks lie east of a line following Slave river and the north arm of Great Slave lake and extending northwest to Great Bear lake. The greater part of the Pre-Cambrian area is underlain by granites and gneisses. Small areas of older, highly metamorphosed rocks are found, such as quartzite, slate, and sericite and chlorite schists; there are also later sandstones that have been intruded by basic igneous rocks. The wide stretches of granite and gneiss offer little attraction to the prospector, but the older, much altered rocks and the later basic dykes, sills, and flows may carry metallic minerals in commercial quantities.

The sediments lying to the west of the Pre-Cambrian area consist of limestones and shales of Devonian age, and shales and sandstones of Cretaceous and Tertiary age, deposited with a great unconformity upon the Pre-Cambrian, but with little unconformity among themselves.

The Devonian limestones and shales form a broad belt stretching from lake Athabaska northward nearly to the Arctic coast. They are concealed to the south and southwest by Cretaceous rocks, and towards the delta of the Mackenzie also pass beneath Cretaceous sediments. These later sediments also overlie the Devonian rocks in long stretches along the Mackenzie above and below Great Bear river and are exposed to the west and north of Great Bear lake. The Devonian rocks are in general nearly horizontal, but at a number of points have been folded into sharp anticlines.

The Cretaceous sediments resting upon the Devonian rocks at the south occupy nearly the whole of the valleys of Athabaska and Peace rivers and probably extend in a northwest direction to Liard river; to the south they stretch beyond the area dealt with in this report through Alberta and Saskatchewan into the United States. They consist of a series of sandstones and shales, and are nearly horizontal in attitude, but dip slightly to the south or southwest in the basins of Athabaska and Peace rivers.

A few small areas of Tertiary sandstones and clays are known. The Paskapoo formation is exposed on Pembina river, a tributary of the Athabaska. There is a small area of Tertiary sediments on the Mackenzie at the mouth of Great Bear river and another on Peel river. These sediments are gently undulating, but do not depart much from the horizontal attitude.

As settlement advances northward and transportation lines are constructed, increased attention is directed to the possible economic mineral deposits of the Mackenzie basin.

Coal seams of workable thickness are found in the Cretaceous sediments and some of these are being mined in the basin of the Athabaska. Lignite seams also occur in the Tertiary beds of Mackenzie and Peel rivers, but these are too remote from a market to be of any importance for many years to come.

Extensive deposits of bituminous sands are exposed on Athabaska river at the base of the Cretaceous system and investigations have been made into their possibilities for commercial application.

A gas-bearing horizon in the Cretaceous, on the Athabaska, was discovered by the Geological Survey twenty years ago. Petroleum has been found in borings on Peace river. Very little of the country, however, has yet been prospected.

Extensive gypsum deposits are found on Peace river in a posi
favourable for quarrying and for transportation. Gypsum is also f
on Slave river, in the escarpment west of Slave river, and elsewhere.

Salt beds have been reported as discovered in bore-holes at McMur
Brine springs west of Fort Smith have furnished that part of the Macke
valley with its salt supply for many decades.

Some clays of good grade are found on the Athabaska.

TABLE OF FORMATIONS.

Quaternary	Pleistocene and Recent			
Tertiary	Paskapoo			
	Not subdivided			
Mesozoic	Cretaceous	Peace River	Edmonton and Wapiti River	Sandstone.
			Upper La Biche and Smoky River	Shale.
			Dunvegan	Sandstone.
			Fort St. John	Shale.
			Peace River	Sandstone.
			Loon River	Shale.
			Lower La Biche	Shale.
		Athabaska River.	Pelican sandstone	
			Pelican shale	
			Grand Rapids	Sandstone.
			Clearwater	Shale.
			McMurray	Sandstone, bituminou
			Not subdivided	Sandstone and shale.

Unconformity.

Palæozoic	Devonian	Limestone and shale.
	Not subdivided	Limestone, dolomit sandstone, and shal

Unconformity.

Pre-Cambrian	Late Pre-Cambrian	Sandstone, limesto and basic flows intrusives.

Unconformity.

	Granites and gneisses

Intrusive Contact.

	Early Pre-Cambrian	Schists, slates, li stones, and qu zites.

DESCRIPTION OF FORMATIONS.

Pre-Cambrian.

The distribution of the Pre-Cambrian rocks of the Mackenzie basin coincides with the physiographic province known as the Laurentian plateau, which occupies a strip along the eastern border of the basin about 800 miles in length and ranging from 80 to 280 miles in width. On the east, the Pre-Cambrian rocks extend far beyond the boundaries of the region over to Hudson bay, but on the west they pass under a cover of later stratified rocks. The contact with these younger rocks runs from Methye portage to the west end of Athabaska lake, thence down the valley of Slave river to Great Slave lake, and from the north arm of that lake north-northwesterly to McTavish bay on Great Bear lake. North of Great Bear lake the contact has not been defined, but it probably soon passes northward outside the watershed of Mackenzie river, reaching the Arctic coast at Darnley bay.

The Pre-Cambrian rocks of this region have not been subdivided in such great detail as they have been in the lake Superior region or northern Ontario, and only three different subdivisions have been recognized and mapped. More detailed work will no doubt show the same complexity of formations as exists in other regions where the Pre-Cambrian has been more carefully studied.

The oldest of the three groups of rocks into which the Pre-Cambrian has been subdivided consists of highly metamorphosed schists, slates, limestones, quartzites, and volcanic rocks, occurring in isolated areas as erosion remnants. This group includes formations that have been uncertainly correlated with the Keewatin, Huronian, or Grenville of other regions.

These rocks have been intruded by granites or gneisses of probably more than one age.

Overlying these two groups is a series of younger rocks, not greatly disturbed, consisting of sandstones, conglomerates, shales, limestones, and some associated volcanic rocks that have been referred to as Keweenawan or Animikie.

Early Pre-Cambrian. Rocks that have been more or less doubtfully correlated with the Keewatin or Huronian and which consist mainly of schists, slates, limestones, and quartzites occur here and there throughout the Pre-Cambrian of the Mackenzie basin in areas usually small but sometimes many square miles in extent. The characteristic feature of the rocks of all the areas is that they are intruded by the granites or gneisses of the succeeding division, though they probably represent formations of vastly different ages. They occur in isolated bands which are merely remnants of formerly more widespread series of rocks that at one time probably covered the greater part of the whole region. They have, however, been so reduced in area by intrusions of granites and by deep erosion that they now form apparently the smallest portion of the Pre-Cambrian of this region.

Three main bands of these rocks on the north shore of Athabaska lake are described by Tyrrell[1] under the name Huronian, and in a later report are referred to by Alcock[2] as portions of the Tazin series.

[1] Geol. Surv., Can., Ann. Rept., vol. VIII, p. 17 D.
[2] Geol. Surv., Can., Sum. Rept., 1914, p. 61.

The largest of these bands is situated on Beaverlodge bay "where it is developed for 16 miles along the shore and extends back at least 10 miles to the north. It consists of limestone, quartzite, slate, and sandstone. The limestone is bluish in colour, weathering to a rusty brown, and occurs only in local patches cut by gneiss. The quartzite is much the most abundant type in the series. It is white and in places reddish, is very badly brecciated, and in several localities contains considerable hematite."[1] The second band on Slate island and the adjacent mainland consists of dark grey and brown schists, quartzite, and breccia. The third band occurs near Sand point and runs parallel to the shore. It consists of beds of quartzite and schist which form cliffs along the lake shore for about 2 miles.

In the country north of Athabaska lake, rocks of the Tazin series are described by Camsell[2] as occurring in seven distinct areas along the Tazin and Taltson rivers. These areas are situated along the canoe route which follows these two streams and are designated as follows: Thluicho Lake area, Thainka Lake area, Long Reach area, Hill Island Lake area, Thekulthili Lake area, Kozo Lake area, and Tsu Lake area. They occur in elongated bands from 6 to 28 miles in length. The rocks strike north or northwest, and consist of schists, quartzites, argillites, limestones, and some volcanic rocks, all dipping at high angles. All these rocks are intruded and completely surrounded by granite or gneiss and have been highly metamorphosed thereby. They generally occupy depressions that form the water-courses. The band on Hill Island lake, which is at least 18 miles in length, is from an economic point of view probably the most important in this region. It consists of narrow, interbedded bands of limestone, argillite, and mica schist, which stand vertically and strike north and south. Numerous veins of quartz mineralized by pyrite cut the beds; veins of the interbedded type are also found. Another band on Thekulthili lake also traversed by quartz veins contains a bed of conglomerate which carries pebbles of an older granite.

On Great Slave lake, rocks consisting mostly of schists which have been classed as Huronian by Robert Bell[3] occur "around Yellowknife bay and thence to Gros Cap, including some islands in this part of the lake, also on islands in the vicinity of Fort Rae and again at the head of lake Marian............Rocks which may belong to this series were observed on the southeast side, to the northeastward of the mouth of Slave river."

Though all these bands are areally of not great extent, they are much the most important economically of all the divisions of the Pre-Cambrian, and, although no commercial mineral deposits have as yet been found in this division within the limits of the Mackenzie basin, the rocks of which they consist have been proved in other parts of Canada and the United States to contain valuable deposits of gold, silver, and iron ores. The rocks are frequently traversed by quartz veins which in the band on Yellowknife river are known to carry free gold. It is advisable, therefore, for the prospector working in the Pre-Cambrian of this region to pay particular attention to these rocks, as they are more likely than any other formation to contain deposits of metallic ores.

[1] Alcock, F. J., op. cit.
[2] Geol. Surv., Can., Mem. 84, p. 25.
[3] Geol. Surv., Can., Ann. Rept., vol. XII. p 106 A.

Granites and Gneisses. Granites and gneisses cover the greater part of the Pre-Cambrian of the Mackenzie basin. Along the canoe route followed by Camsell between Athabaska and Great Slave lakes by way of the Tazin and Taltson rivers, granites and gneisses "occupy about 86 per cent of the region immediately adjacent to the line of traverse, and it is very likely that they cover by far the greater part of the whole Taltson River basin, because most of the glacial boulders scattered over the surface of the region are of these rocks."[1] Although this proportion may not hold over the whole of the Pre-Cambrian it is cited as evidence of the great areal extent of the granites and gneisses relative to the other rocks of this age.

On Athabaska lake Tyrrell[2] describes the rocks of this group as consisting of hornblende granites, biotite granites, muscovite granites, and granitoid gneisses with gabbros and norites. They are all closely associated with each other and although several ages are represented they are all classed together in default of evidence whereby they could be arranged in a definite time series.

The outcrops are here, as elsewhere, worn into rounded hills and ridges. The gneiss does not appear to have a persistent strike, though Alcock notes a tendency to a northeasterly trend. Under the microscope it exhibits a cataclastic structure indicating subjection at some time to severe pressure.

North of Athabaska lake, in the basin of Taltson river, the same types of rocks prevail with the addition of some diorites. In this region Camsell[3] notes a trend of the foliation of the gneiss ranging from north to northwest, coinciding with the trends of the bands of stratified rocks into which the gneiss is intruded. The trend here is referred to as a Cordilleran one rather than one conforming to the trend of Pre-Cambrian rocks in eastern Canada.

In the Great Slave Lake region and northward to Great Bear lake, granites and gneisses constitute a very large proportion of the bedrock. Porphyries, syenites, and granites are described by J. M. Bell[4] as occurring all the way from Great Bear lake to Great Slave lake, and on the east arm of Great Slave lake "granites and gneisses rise as a sea of half rounded hummocks to a general height of nearly 1,000 feet all along the northwest side of this part of the lake and also around the northeastern extremity."[5]

Associated with the granites and gneisses, but possibly of a later period of intrusion, are bodies of gabbro and norite occurring notably on the north shore of Athabaska lake and the country to the north.

This whole group of rocks belongs not to one period of intrusion alone, but represents different intrusions. In the main, however, the rocks cut the series of stratified rocks locally known as the Tazin series and others that have been referred to as Keewatin or Huronian. They are not in themselves mineralized to the extent of forming workable ore-bodies, but they are in places believed to have produced mineralization in the rocks into which they have been intruded.

Late Pre-Cambrian. The youngest Pre-Cambrian formations of this part of Canada consist chiefly of sandstones associated with basic intrusions

[1] Camsell, Charles, Geol. Surv., Can., Mem. 84, p. 33.
[2] Tyrrell, J. B., Geol. Surv., Can., Ann. Rept., vol. VIII, p. 16 D.
[3] Camsell, Charles, Geol. Surv., Can., Mem. 84, p. 33.
[4] Geol. Surv., Can., Ann. Rept., vol. XIII, p. 101 A.
[5] Bell, Robert, Geol. Surv., Can., Ann. Rept., vol. ""I, p. 106 A.

and flows. These rocks are of wide distribution; they lie to the north-east of Great Bear lake, at the east end of Great Slave lake, and north and south of lake Athabaska; they also cover large areas lying to the east of the Mackenzie basin.

In the Lake Athabaska district the rocks consist chiefly of reddish, moderately coarse-grained sandstones, becoming in places fine-grained, thin-bedded, and shaly, and in places passing downward into conglomerate. These constitute what have been designated the Athabaska series.

[1] "The total area underlain by this sandstone formation is very large, extending from Cree lake on the south to Athabaska lake on the north, and from Wollaston lake on the east, doubtless to the vicinity of the valley of Athabaska river on the west, and perhaps much further under the covering of later rocks. Cree lake lies largely within the area underlain by these rocks, and Athabaska lake seems to lie entirely within it, for the red sandstones compose many of the islands and more prominent points of its northern shore."

North of Beaverlodge bay on the north shore of lake Athabaska, thick-bedded sandstones, arkoses, and conglomerates of the Athabaska series are found, and interbedded with these is a number of flows of vesicular and amygdaloidal basalt. The series is also intruded by sills and dykes of diabase. The strata here form an open syncline and have a maximum dip of 40 degrees. There is a thickness of about 8,800 feet.[2] Conglomerate and shaly sandstone exposed at the northeast end of Tazin lake possibly belong to the same series.[3]

Concerning the formation as exposed on Great Slave lake, Robert Bell says:

[4] "The northeastern continuation of the main lake-basin is excavated out of the older Cambrian or Animikie rocks resting in a long physical depression or trough in the Archæan foundation. These strata have a thickness of over 1,000 feet and they are thrown into gentle anticlines and synclines, parallel to the axis of the general trough, in which they lie. They have been deeply eroded along the anticlinal folds and the waters now filling the depressions form the various long and nearly parallel bays into which this portion of the lake is divided. These rocks consist partly of unaltered limestones varying in colour from very light- to dark-grey, drab, and red, sometimes passing into shales, and partly of sandstones, mostly red, coarse conglomerates, and red shales, together with thick sheets or overflows of greenstone, generally capping the other strata and presenting long cliffs made up of perpendicular columns or "palisades' overlooking the different bays. We could not ascertain whether all these greenstone cappings belonged to a single extensive overflow or not. Large exposures of greenstone also occur near the level of the lake, which may not form part of any general overflow. A few wide greenstone dyke were seen cutting the nearly horizontal Animikie strata beneath the crowning overflow."

In the narrows southeast of Big Caribou island and on the tongue of land separating McLeod bay from the east bay, there are exposure of "massive light-grey, blue, or dove-coloured limestones which weathe

[1] Tyrrell, J. B., Geol. Surv., Can., Ann. Rept., vol. VIII, p. 18 D.
[2] Alcock, F. J., Geol. Surv., Can., Sum. Rept., 1916.
[3] Camsell, Charles, Geol. Surv., Can., Mem. 84, p. 36.
[4] Geol. Surv., Can., Ann. Rept., vol. XII, p. 106 A.

to various shades of yellow and brown, hard reddish sandstones or quartzites and fine conglomerates, and red and grey 'lumpy' jasper or chert-rock. At the cast bay, black shale occurs in the vicinity of the massive limestone."

Sedimentary rocks of various kinds, associated with basic igneous intrusives, are exposed at the eastern end of Great Bear lake and extend eastward beyond Coppermine river and northward to the Arctic ocean.

At Limestone point, on the north shore of Dease bay, 30 miles west of Fort Confidence, there is an exposure of " purplish dolomite which changes to a ferruginous slate. Above this comes grey, semi-crystalline dolomite, associated with light-grey quartzite." Limestones and slates are exposed at points along the shore to the east of this. Exposures of limestone, quartzite, sandstone, and basic igneous rocks occur on the rough, rocky south shore of Dease bay as far west as Narrakazoo islands. On these islands greenstone is exposed. West of this the south shore of Dease bay is low, as is also the north shore of McTavish bay for 50 miles east of cape McDonnel. Farther east the shore of McTavish bay is rocky, and greenstone intrusions are found cutting horizontal sediments.[1] " The eastern part of McTavish bay is composed of a series of basic rocks, or greenstones, that seem to overlie the Laurentian granites, of which, however, exposures are seen at several places. The southern part of McTavish bay and the islands there, are mostly of granite, though greenstone dykes are common." On Dease river bright red quartzite, and drab and red magnesian limestones occur, and at its mouth exposures of diabase and diorite are seen.

On Coppermine river, which lies east of Great Bear lake outside the basin of the Mackenzie, a number of basaltic flows, the upper parts of which are amygdaloidal, are found interbedded with conglomerate, sandstone, and shale. These rocks are of wide distribution, being found on Coppermine river, on the coast and islands to the north, and on Bathurst inlet. Their chief interest lies in the fact that they carry native copper.[2]

Palæozoic (not subdivided).

A range of mountains crosses Great Bear river at the rapids and extends northwest and southeast beyond the range of vision. At Great Bear river the highest point is mount Charles, which has an elevation of about 1,500 feet. The culminating peak of the range to the southeast is mount Clark; this has a height of 3,000 to 4,000 feet. These mountains run nearly parallel to Mackenzie river as far south as Willow river and lie a few miles to the east of the Mackenzie. Mount Charles consists of interstratified conglomerates, quartzites, and magnesian limestones of Palæozoic age.[3] Good sections of Palæozoic sediments are exposed in the Rocky and Mackenzie mountains.

[1] Bell, J. M., Geol. Surv., Can., Ann. Rept., vol. XII, p. 26 C.
[2] Douglas, James, "The copper-bearing traps of the Coppermine river," Can. Min. Inst., Trans., vol. XVI, pp. 83-101.
O'Neill, J. J., Geol. Surv., Can., Sum. Rept., 1916.
[3] Bell, J. M., Geol. Surv., Can., Ann. Rept., vol. XII, p. 25C.

Devonian.

In western Canada the boundary between the Pre-Cambrian formations of the Canadian Shield to the east and the Palæozoic formations of the prairie provinces and the Northwest Territories to the west lies approximately along a line joining lakes Winnipeg, Athabaska, Great Slave, and Great Bear. Of the Palæozoic systems of rocks lying immediately to the west of this line, the Devonian is in point of areal extent the most important. Future investigations may also prove it to be the most important from an economic standpoint.

Insufficient work has been done to determine the exact boundaries of the Devonian rocks. The eastern border lies approximately along Slave river and the north arm of Great Slave lake and then swings somewhat to the west of the chain of lakes and rivers forming the canoe route between the north arm of Great Slave lake and the east end of Great Bear lake. To the south and west the Devonian rocks pass beneath sediments of Cretaceous age.

The Devonian limestone extends south of the west end of lake Athabaska and outcrops in long narrow bands in the bottoms of the valleys of Athabaska river (Plate X) and its tributaries where these have cut their way through the overlying Cretaceous. It stretches up the Athabaska beyond McMurray to a point 2 miles above Crooked rapid. It is exposed on Firebag, Muskeg, McKay, and Clearwater rivers and reaches several miles up Christina river.

On the Athabaska the limestone is evenly-bedded, greyish in colour, somewhat argillaceous, and passes in places into calcareous shale. The strata are in general horizontally disposed, but there are numerous gentle undulations.

The boundary between the Cretaceous and the Devonian runs to the west of the lower Athabaska, swings around the north end of Birch mountain, extends westward, and crosses Peace river above Vermilion falls. Devonian limestone is exposed in Hay river for a distance of 3 miles above Alexandra falls. The 700 or 800-foot escarpment lying 30 miles north of lake Bistchô may be composed of Cretaceous rocks.

Exposures occur on Steepbank river, which flows into lake Claire and on the lower 2 miles of Mikkwa river, which empties into the Peace below Vermilion falls. Cream-coloured limestone is brought to the surface by an anticline between 35 and 40 miles up Mikkwa river. This is at a point 6 miles above the exposures of Cretaceous shale.

On Peace river Devonian limestones are exposed in low cliffs from the mouth of Mikkwa river to 2 miles above Vermilion falls, and in the vicinity of the falls have an exposed thickness of 60 feet.[1] The limestone at this point is horizontal and consists of thick, evenly-stratified, light greyish or cream-coloured beds, alternating with softer and more argillaceous bands. Devonian strata are also exposed in cliffs 20 to 60 feet high between Little rapids and Peace point. The rocks consist of gypsum[2] overlain by 10 to 30 feet of fractured and broken dolomite which is succeeded upward by argillaceous limestone.[3]

[1] McConnell, R. G., Geol. Surv., Can., Ann. Rept., vol. 5, p. 45D.
[2] Palæontological evidence recently obtained by E. M. Kindle indicates that these and other beds of gypsum throughout the basin of Mackenzie river are of upper Silurian age.
[3] Macoun, John, Geol. Surv., Can., Rept. of Prog., 1875-76, p. 89.
Camsell, Charles, Geol. Surv., Can., Sum. Rept., 1916.

A number of exposures of Devonian strata occur on Slave river. On the west bank, at La Butte, there is a section, in ascending order, of 10 feet of thin-bedded, grey limestone, 6 feet of brecciated limestone impregnated with bitumen, 10 feet of pebbly limestone, and 5 feet of massive limestone. A few miles below La Butte there is a 10-foot bed of impure gypsum overlain by 20 feet of fractured and broken limestone. A number of good exposures occur between Stony islands and Caribou island. At Bell rock, 7 miles below Fort Smith, a brecciated limestone is exposed, and on the east side of the river immediately below point Ennuyeux there is an exposure of 4 feet of gypsum overlain by shaly limestone.[1]

Limestone and gypsum beds are exposed in the escarpment lying a few miles west of Slave river. This escarpment runs northwest and southeast; at the brine springs it is 210 feet high and at Little Buffalo river it is 150 feet high. No exposures have been observed on Jackfish river, though large, angular blocks of limestone are common a few miles below the sulphur springs.[2]

The west arm of Great Slave lake rests on flat-lying Devonian limestones. Exposures occur along the shore at a number of points and in other places numerous angular blocks are found that have not travelled far. In the vicinity of Resolution, and on Mission and Moose islands, angular blocks are numerous and in places dark-grey, thin-bedded limestones in place are exposed. Lime tone outcrops at a few places west of the mouth of Little Buffalo river, and highly bituminous, calcareous shales and thin-bedded limestones outcrop on the shore in the vicinity of Pine point and on the islands east of the point. An anticlinal structure is observed on the shore at this point. Seven miles southeast of Pine point a flat-topped hill 200 feet high, situated $1\frac{1}{2}$ miles inland, is capped by a heavy-bedded, soft, yellow, calcareous sandstone. At Presqu'ile some of the exposed limestone beds are porous and carry bitumen. Exposures of limestone occur at Sulphur point, but west of that no outcrops have been observed on the south shore of the lake although angular blocks are found.

On the north shore of Great Slave lake outcrops occur between Slave point and the north arm of the lake. On Slave point bituminous limestone is found, and on Windy point a 15-foot cliff of limestone and shale extends 3 miles in a northerly direction. The northeast portion of Windy point is composed of dolomite giving seepages of petroleum, described elsewhere in this report. On the north shore of the deep bay lying to the north of this point there is a range of hills composed of dolomite, and tar springs occur about a mile inland. East of this bay bituminous limestone outcrops at nearly all points, stretching eastward into the lake. At Gypsum point and along the southwest shore of the north arm of the lake calcareous sandstone and arenaceous limestone outcrop at many points and carry thin seams of gypsum in the bedding planes.[3] The boundary between the Palæozoic and the Pre-Cambrian lies about 2 miles east of Fort Rae.

Good sections of the Devonian rocks are exposed in the valley of Hay river (Plate XIA), at Alexandra falls, in the gorge below the falls, and in

[1] Camsell, Charles, Geol. Surv., Can., Sum. Rept., 1916.
[2] Camsell, Charles, Geol. Surv., Can., Ann. Rept., vol. XV, p. 158A.
[3] Cameron, A. E., Geol. Surv., Can., Sum. Rept., 1916.
    The Geology of Hay river and of the country about the western part of Great Slave lake, as laid down on the map accompanying this report, is taken from the manuscript map of A. E. Cameron.

bluffs farther down the river. The following geological section was mea
sured by A. E. Cameron:

	Feet.
Massive, arenaceous limestone	30
Thick and thin-bedded limestones, shaly bands with fossils, fine-grained, grey	62
Coralliferous limestone, bituminous	16
Red-brown, medium-grained limestone	12
Blue-grey shale	5
Reddish sandstone, ripple-marked	7
Blue-grey, soft shales	47
Massive, red-brown limestone	8
Shaly limestone, many fossils	12
Thin-bedded, brown limestone	21
Blue-green, clay shales, limestone layers with many fossils	42
Highly fossiliferous limestone	20
Blue-green, clay shale	15
Flaggy sandstone, shaly layers with ripple-marks and worms	14
Fossiliferous limestone	8
Blue-grey, clay shale	28
Thin-bedded, fossiliferous limestone	8
Blue-green, clay shales, limestone bands with fossils	105
Thin-bedded limestone, fossils	8
Blue-green, clay shales	10
Sandstone, ripple-marks and worms	12
Blue-green, clay shales, limestone bands with fossils	90 aneroid.
Thin-bedded limestone, shaly, argillaceous	25
Blue-green, clay shales, bottom not exposed	

605

The strata on Hay river appear to have a slight dip upstream an
around Great Slave lake there appears to be a slight dip to the wes'
although in all cases where thin-bedded limestones outcrop at watel
level the dip is apparently toward the lake. Small local domes with
diameter of 100 to 200 feet and with dips seldom exceeding 5 degrees ar
common.

The following notes by A. E. Cameron on the Devonian formation
exposed in the vicinity of Great Slave lake are taken from the Summar
Report of the Geological Survey for 1917, which was published since th
report was compiled.

"*Lower Devonian.* No distinctly lower Devonian sediments wer
observed in the vicinity of the lake shores or on Peace river. Kindl
reports shales carrying middle Devonian fossils suggestive of the Ithac
formation, a local facies of the Portage in New York directly overlyin
the Gypsum series on Peace river. An erosional unconformity exists hei
between the Gypsum series and the overlying shales. On Great Slav
lake, however, a thick series of middle Devonian sediments are foun
lying between the Simpson shales and the upper Silurian dolomitic lim
stone. Fossil evidence shows that the Simpson shales are equivalent t
the Portage formation of New York.

"*Middle Devonian.* Middle Devonian sediments exposed on th
shores of the lake are divided on lithological and palæontological evidenc
into three formations: Slave Point limestones, Presqu'ile dolomites, an
Pine Point limestones. These may be correlated with the Manitoba
limestones, Winnipegosan dolomites, and Elm Point limestones of th
Manitoba section.

"The Pine Point limestones, the lowest member of the series, ai
exposed in the vicinity of Resolution, at Pine point on the south shore

the lake, and on Ketsieta point on the north shore. They are thin-bedded, bituminous, dark-coloured, fine-grained limestones and tiny shales.

" The thickness of the Pine Point series is not directly observable, but from structural evidence appears to be about 100 feet.

" The Presqu'ile dolomites, overlying the Pine Point series, are exposed at Presqu'ile point and on the Burnt islands east of Pine point on the south shore. On the north shore they show as the oil-bearing dolomites at the tar springs on Nintsi (Windy) point and on the shores of Sulphur bay. Although not exposed elsewhere on the north shore of the lake, structural conditions suggest their presence in the region between Tagkatea and Ketsieta points.

" Exposures show this formation to consist of two members: an upper thin-bedded dolomitic limestone highly fossiliferous and carrying the diagnostic fossil *Stringocephalus Burtoni*, and a lower member composed of a coarsely crystalline porous and cavernous dolomite. This latter is the oil-bearing horizon at the tar springs on Nintsi point. The formation is estimated to have a thickness of about 200 feet.

" The Slave Point limestones, composing the upper formation of the middle Devonian, are exposed on the south shore, from Presqu'ile point to High point, on Buffalo river, and on the north shore at Slave point and along the lake shore between House and Moraine points.

" This formation is composed of thin-bedded, medium-grained, dark grey, and slightly bituminous limestones and has an estimated thickness of about 160 feet.

" *Upper Devonian.* Upper Devonian sediments in the region are divided into three formations: Hay River limestones, Hay River shales, and Simpson shales. The Hay River limestones and shales carry an abundant *Spirifera disjuncta* fauna and may be correlated with the Chemung formation of New York. and the Simpson shales carry some of the fossils characteristic of the Portage formation of New York.

" The Simpson shales are not exposed in the vicinity of Great Slave lake, but on Mackenzie river near Simpson are found underlying the Hay River shales. A thickness of 150 feet is exposed and shows soft, greenish grey clay shales. Limestone and sandstone bands, so common to the Hay River shales, are absent here, and the fossil evidence does not show the *Spirifera disjuncta* fauna which is so abundant in the Hay River series. It is probable that these soft Simpson shales underlie the western end of the lake and a considerable portion of the valley of Buffalo river.

" The Hay River shales are exposed in the valley of Hay river below the falls and are also probably present underlying the basin of Buffalo lake and in the valley of Beaver river. Where exposed on Hay river they show as soft, bluish-green, clay shales and carry thin bands of highly fossiliferous limestone and ripple-marked sandstones. A measured thickness of 400 feet of these shales is exposed in the river valley.

" The Hay River limestones directly overlie the soft Hay River shales and the section as exposed in the gorge on Hay river shows a gradation between the two formations.

" The section in the gorge on Hay river shows 221 feet of thick and thin-bedded, hard, fine-grained, light-coloured limestones and at least another 75 feet is exposed in the river valley above the falls. About 40 feet from the base of the limestones occurs a bed of soft fossil clay shales 47 feet thick. The presence of this shaly member in the limestone

series has been the cause of the formation of the two falls on Hay rive
At the falls on Beaver river the limestones are similar to those at Loui
falls on Hay river. The shaly member is here absent and in place of a
upper fall there is a long series of low cascades from 5 to 15 feet high."

Horizontal, compact limestone was observed by McConnell ju
south of lake Kakisa, between Providence and lake Bistchô.

Numerous outcrops occur along Mackenzie river. Bluish shales a
exposed at a number of points between Providence and Simpson. Hor
mountain and Trout mountain, which lie respectively to the east an
west of the river, may be escarpments formed by limestone beds overlyin
these blue shales.[1]

For a few miles up Liard river there are no outcrops, but above th
comes a stretch of 25 miles in which the river flows between steep banl
200 to 400 feet high, composed of Devonian shales, limestones, and ca
careous sandstones lying in a nearly horizontal position. Limeston
form the top of the section. Limestone is also exposed at points 16 an
24 miles below Fort Liard.

Shales outcrop again in a high hill on the right bank of the Mackenz
4 miles below Willow river, and thick-bedded, dark-grey limestone is see
on two islands opposite old Fort Wrigley. Eleven miles below old Fo
Wrigley Devonian shales are exposed, and in the "Rock-by-the-rive
side," which rises with a steep face towards the river to a height of 1,50
feet, Devonian limestones dip to the west at an angle of between 60 and
degrees. Six miles below the "Rock-by-the-river-side" horizontal, greyis
shales outcrop, and 6 miles farther downstream they appear again an
recur at intervals for several miles. Ten miles above Blackwater riv
the plateau lying 3 or 4 miles east of the Mackenzie and facing westwar
rises to a height of 1,000 feet above the plain, and in it westerly dippin
Devonian limestones are exposed.

Devonian limestones and shales outcrop again for a few miles belo
the mouth of Great Bear river, and Bear mountain, which rises to a heigl
of 1,400 feet just below Great Bear river, is composed of limestones, quart
ites, and shales folded in an anticline. "The lowest beds seen consi
of reddish and greenish shales, alternating with layers of pink-coloure
gypsum, and cut by numerous veins and seams of a white fibrous variet
of the same mineral. The gypsum in parts of the section replaces tl
shales almost altogether, and the layers are separated by mere films
greenish and reddish argillaceous material. The base of the gypsiferou
shales was not seen, but they are at least several hundred feet in thicknes
They are overlain by a series of dolomites, quartzites, and limestone
six to seven hundred feet thick, and then by the bluish coral-bearing lim
stones of the Devonian. Some of the limestone is bituminous and emits
fetid odour when struck."[2]

Fifteen miles below Great Bear river Devonian shales are exposed c
the left bank of the Mackenzie, and in the next 25 miles small sections
bluish shale, shaly sandstones, and limestones holding Devonian fossils a
exposed at intervals. In Roche Carcajou and in the East-mountain-of-tl
rapid, Devonian limestone is exposed in anticlinal uplifts and the Wes
mountain-of-the-rapid is probably of the same composition. At Beave
tail point, limestones dip to the west at an angle of 15 degrees.

[1] McConnell, R. G., Geol. Surv., Can., Ann. Rept., vol. IV, p. 85D.
[2] McConnell, R. G., Geol Surv., Can., Ann. Rept., vol. IV, p. 101D.

At the Ramparts (Plate XIB), which lie a few miles above Good Hope, the river flows for 3 or 4 miles through a gorge 500 yards wide, having vertical walls of Devonian limestone and shale. At the upper end of the gorge the cliffs are 125 feet high; they increase towards the lower end to a height of 250 feet. "The limestone undulates at low angles, but also rises steadily in the walls of the cañon as we descend, at the rate of about fifty feet to the mile. In the upper part of the cañon the walls are precipitous and consist of limestone throughout. The limestones are generally granular in texture and weather to a light cream colour. Some of the upper beds are brecciated and lower down in the band a large proportion of the mass of the rock consists of various species of corals. Thin beds of shale, attenuated to mere films in some instances, are interstratified with the limestone, and increase in importance towards the base. This band of limestone has a thickness of between 150 and 200 feet. Part way down the cañon, bluish green shales holding beds of limestone at intervals, appear below the limestone band, and increasing gradually in height form the basal part of the walls of the cañon the rest of the way. The banks here weather into a steep slope below, but are crowned with almost vertical cliffs above. These shales are precisely similar to those found in the Liard or Hay river, and underlie a similar limestone. The fossil evidence demonstrates that they occupy the same horizon in the Devonian series. This close resemblance is somewhat remarkable when we bear in mind that the two localities are separated by a distance of five hundred and seventy miles."

Three miles below Hare Indian river, Devonian shales are exposed on the right bank of the Mackenzie, and 5 miles farther down, shales, which are evidently the same as those found at the Ramparts, are found on the left bank overlain by yellowish limestone. "In the next twenty-five miles, or as far as the Grand View, Devonian shales overlain occasionally by yellowish-weathering fossiliferous limestones are exposed at intervals all along. The left bank of the river on this reach is usually low and sloping, but the right is bounded for some miles by high limestone cliffs similar to those at the Ramparts. At the head of the Grand View the cliffs leave the river and bend more to the north." Along the Grand View dark argillaceous shales, some of which are bituminous, and dark greyish, sandy shales outcrop at the water's edge. They evidently belong to the same horizon as those at the Ramparts.

Fifteen miles below the Grand View, Devonian shales associated with some sandstone and shaly limestone occur in a small plateau situated some distance east of the river. The shales are highly charged with bitumen, and display a red coloration on the surface which is probably due to combustion of the bituminous matter. Fifteen miles farther down, black, evenly bedded, highly bituminous Devonian shales are found. "The laminæ, when freshly separated, are moistened on the surface with an oily liquid, and burn when thrown into the fire, and patches of red shales, marking the sites of former fires, alternate with the dark varieties. The shales are exposed in the right bank for some miles, or almost as far as old Fort Good Hope." Here they disappear, and Cretaceous sediments outcrop 20 miles below old Fort Good Hope and at other points lower down.

Little is known about the east and west extension of the Devonian system in the lower part of the Mackenzie basin. At Willow river a range

[1] McConnell, R. G., Geol. Surv., Can., Ann. Rept., vol. IV, p. 106D.

of hills appears to the east of the Mackenzie and runs parallel to it at a distance of about 10 miles, crossing Bear river at the rapids and extending many miles in a northwest by north direction. Mount Charles on Great Bear river is about 1,500 feet high and consists of conglomerates, quartzites and a great thickness of magnesian limestone and thin layers of gypsun brought up in a large anticline. These rocks may be of Ordovician o Silurian age.[1] The Devonian system, however, in spite of this break reaches to the south shore of Great Bear lake, to the line of lakes and streams between McTavish bay and the north arm of Great Slave lake at one or two points, such as Mazenod and Marian lakes, and nearly to this line at other points. Fossils of Devonian age were picked up by R MacFarlane on the west side of Rorey lake, but the rock was not found in situ.[2] Fossiliferous limestone was also seen by MacFarlane on Anderson river above the mouth of the Lockhart, one of its tributaries, and fossil. that he collected on the Anderson are of Devonian age.[3]

A cliff of Devonian limestone exceeding 100 feet in height occurs on Gull lake cast of the Mackenzie delta.[4] Gull lake is probably Campbel lake, or one of the small lakes near it lying between the Mackenzie delta and the most southerly of the Eskimo lakes.

The following description by McConnell summarizes much of our knowledge of the lithological character of the sediments of the Devonian system.

"Throughout the Mackenzie district the Devonian is generally divisible lithologically into an upper and lower limestone, separated by a varying thickness of shales and shaly limestones, but in some cases lime stones occur throughout. The upper division has an approximate thickness of 300 feet and consists of a compact, yellowish weathering limestone occasionally almost wholly composed of corals, interstratified with some dolomite beds. This limestone is well exposed at the falls on Hay river and also at the Ramparts on the Mackenzie. In both these places it is underlain by several hundred feet of greenish and bluish shales, alternating with thin limestone beds. At the "Grand View" on the Mackenzie the shales are hard and fissile, and are blackened and in places saturated with petroleum. At the "Rock-by-the-river-side," and at other places where the beds are tilted and older rocks exposed, the middle division is underlain by 2,000 feet or more of greyish limestones and dolomites interbedded occasionally with some quartzites. No fossils were collected from the lower part of this series, and rocks older than the Devonian may possibly be represented in it."

The Devonian strata exhibit a remarkable horizontality throughout the Mackenzie basin. Although minor undulations (Plate X) occur on the Athabaska and its tributaries, and the strata around the west end of Great Slave lake and on Hay river appear to dip slightly to the west or southwest, there is evidence that the shales exposed on Hay river and at the Ramparts are of about the same geological horizon notwithstanding the fact that the two points are separated by a distance of 570 miles.[5]

[1] Bell, J. M., Geol. Surv., Can., Ann. Rept., vol. XII, p. 25C.
[2] MacFarlane, R., "On an expedition down the Beghula or Anderson river," Can. Rec. Sc., vol. IV, p. 30.
[3] Meek, F. B., "Remarks on the geology of the valley of Mackenzie river, with figures and descriptions o fossils from that region, in the museum of the Smithsonian Institution, chiefly collected by the late Rober Kennicott, Esq." Chicago Acad. Sc., Trans., vol. I, pp. 61-114, 1869.
[4] Kindle, E. M., Am. Jour. Sc., vol. XLII, p. 246, Sept., 1916.
[5] McConnell, R. G., Geol. Surv., Can., Ann. Rept., vol. IV, p. 15 D.

Striking breaks in this horizontality are found and at a number of places the sediments have been folded so that the beds are highly tilted. Examples of this folding are seen in the mountain crossing Great Bear river, in the "Rock-by-the-river-side," in the plateau where examined 3 or 4 miles east of the Mackenzie at a point 10 miles above Blackwater river, in Bear rock, in Roche Carcajou, in the East-mountain-of-the-rapid, and at Beavertail point.

Cretaceous.

Sandstones and shales of Cretaceous age are of wide distribution in the southern and northern parts of the region drained by the Mackenzie.

In the southern part they extend from the Rocky mountains east to beyond the Athabaska, being exposed for many miles up Clearwater river and other tributaries from the east. Southward they reach beyond the 49th parallel of latitude. Their northern boundary crosses Athabaska river a little above Firebag river and crosses the Peace a short distance above Vermilion falls. They are cut through to the underlying Devonian limestone by Hay river at a point 33 miles above Alexandra falls. The 700- to 800-foot escarpment lying 30 miles north of lake Bistchö may be composed of Cretaceous rocks. The formation as exposed on Liard river extends from a point a few miles below Fort Liard to many miles above the mouth of Fort Nelson river. On the Fort Nelson and its tributary, the Sikanni Chief river, shales and sandstones occur.

In the northern part of the Mackenzie basin Cretaceous rocks occur in long stretches along the river below the mouth of Dahadinni river, along the upper course of Great Bear river, and around the west and north sides of Great Bear lake. The Cretaceous rocks outcrop also in the Mackenzie below old Fort Good Hope and in the basin of Peel river.

On Athabaska river the following succession of sediments occurs:

Edmonton formation.
La Biche shales.
Pelican sandstone.
Pelican shale.
Grand Rapids formation.
Clearwater formation.
McMurray sandstone.

The sediments are nearly horizontal, dipping to the south only a few feet a mile, so that in descending the river one finds each formation emerging from beneath the overlying one in the order named in the table.

The McMurray sandstone is prevailingly arenaceous and is rather coarse-grained. Large portions are impregnated with bituminous matter (Plate XII A and B) forming the so-called tar sands or bituminous sands. The formation is exposed along the Athabaska from Boiler rapid to below Calumet river and along the valleys of tributary streams. It has a thickness of 110 to 180 feet. The top of the formation is placed at the base of a bed of green sandstone immediately above which the marine fauna of the Clearwater formation appears. It contains lignite seams and fragments of fossil wood and carries a small invertebrate freshwater fauna, but its exact age has not been established.[1]

_____
[1] McLearn, F. H., Geol. Surv., Can., Sum. Rept., 1916.

The Clearwater formation consists of soft grey and black shales grey and green sandstones and contains some hard concretionary lay It has a thickness of 275 feet, first appears 5 miles below Grand island, and extends down stream above the McMurray sandstone. It carries a mar fauna.

The Grand Rapids formation is a sandstone 280 feet thick. It first exposed 3 miles above Joli Fou rapid, and below Grand rapids for almost continuous cliffs for miles. The upper part carries thin coal sea and is of subaerial origin; the lower part is highly concretionary, and large concretions carry a marine fauna.

The Pelican shale, which has a thickness of 90 to 100 feet, is a bla shale that first appears near Stony rapids and is a conspicuous feature many miles down the Athabaska. It thins out toward the northw being only a few feet thick on Moose river and not recognized on the no end of Birch mountain. It carries poorly preserved species of *Inoceram*

The Pelican sandstone appears from beneath the La Biche shales the mouth of Pelican river and forms a low cliff in all the exposures betwe Stony rapids and Grand rapids. It is 35 feet thick and is in general co picuously white. A species of *Inoceramus* is found at the top.

The La Biche shales overlie the Pelican sandstone and extend up t Athabaska a few miles beyond the mouth of Pembina river, and up Les. Slave river into the country surrounding Lesser Slave lake. The sha are also exposed in Birch mountain, an erosion plateau lying to the w of the lower part of Athabaska river. A section at Athabaska, as expos in the bank and revealed by boring, consists of 15 feet of yellowish sar stone at the top, 165 feet not well exposed, and 1,090 feet of grey and bla ish shales. These are succeeded by Pelican sandstone.[1] The few fos. in the upper part of the formation are typical Pierre; the lower part probably older than the Pierre[2].

The Edmonton formation consists of yellowish and greyish, flaggy a massive sandstones, often holding large concretions, and alternating wi greyish and dark clays and shales. These sandstones, clays, and sha carry seams of lignite and are practically horizontal. The formati extends in a broad belt eastward from the foothills nearly to the juncti of the Pembina and the Athabaska. Sediments that may belong to t formation are well exposed in the plateaus south of Lesser Slave lake whe they attain a thickness of at least 1,000 feet.[3] Dowling has suggested tl these are equivalent to the top of the Belly River formation.[4]

Marten mountain, which lies northeast of Lesser Slave lake and 1,000 feet high, is composed largely of sediments that may be equivale to the Edmonton in age. Fragments of lignite and sandstone a abundant on its lower slopes and a loose fragment of sandstone found its base afforded specimens of freshwater shells.

The Cretaceous sediments of Athabaska river are nearly horizont and rest upon the Devonian with an unconformity so slight that it can detected only by examination of the contact for a distance of many mil Between Grand rapids and Pelican rapids the dip is about 5½ feet per m south. Between Pelican rapids and Athabaska the dip is about 10 feet j

[1] Dawson, G. M., Geol. Surv., Can., Ann. Rept., vol. XII, p. 14A.
[2] McConnell, R. G., Geol. Surv., Can., Ann. Rept., vol. V, p. 28D.
[3] McConnell, R. G., Geol. Surv., Can., Ann. Rept., vol. V, p. 53D.
[4] Roy. Soc. Can., Trans., ser. 3, vol. 9, sec. 4, p. 39.

mile. This greater dip may be due, in part, to its southwest direction, and may indicate that the true dip of the Cretaceous strata on the Athabaska is west of south. A low anticline crosses the river near Crooked rapids; the dip on each side is only 3 or 4 feet a mile. Below McMurray the strata are probably nearly horizontal and may have a slight north or northeast dip.

The Cretaceous section exposed on Peace river is described as follows by F. H. McLearn in the Summary Report of the Geological Survey for 1917.

" The rocks exposed on Peace river may be resolved into an eastern and western succession,. the former ascending from Vermilion chutes to Dunvegan and the latter descending from there to the mountains. Characteristics of the succession are: the presence of a marine Cretaceous fauna prior to, and quite unlike that of, the Colorado; subaerial delta deposits are present in the Colorado group.

Eastern Succession.

" *Devonian Limestone.* Limestones of Devonian age outcrop from Vermilion chutes downstream.

" *Loon River Formation.* The Loon River shales are exposed on the valley sides from Vermilion chutes to near Brown's trading post, and north and east of the great horseshoe bend underlie the plateaus adjacent to the river. They consist of dark blue to dark grey, friable weathering shale with a few rounded or flattened ironstone concretions. To the south, where they are penetrated by the wells of the Peace River Oil Company, they are more arenaceous, particularly near the base. This, in well No. 1 the following section rests on the limestone.

	Feet.
Shale	70
Sandstone	53
Shale	65
Sandstone	12
Shale	26
Sandstone	51
Shale with concretions and bands of ironstone	
Palæozoic limestone	—

" In No. 2 well, about 1½ miles south of No. 1, the 70-foot bed of sandstone has increased to 106 feet and the 53-foot bed of shale has decreased to 14 feet. Both wells also show a number of smaller beds of sandstone in the upper part of this formation, which are not present in the exposures downstream. Along the upper contact with the Peace River sandstones there is much replacement laterally of sandstone by shale, so that the contact rises stratigraphically northward. As exposed on the river in the north the even bedding and marine fossils point to a marine origin. It is possible, however, that the thick sandstones near the base of the formation in the south may be of non-marine origin, like the tar sands of the Athabaska at the same horizon. At the great horseshoe bend the thickness is 400 feet (estimated) and to the south in the oil wells 1,100 feet.

" *Peace River Formation.* The rocks of the Peace River formation outcrop on both sides of the valley from a point about 14 miles above Carcajou point to Peace river. Where they are typically developed and exposed, they form steep valley-walls, the " Ramparts of the Peace."

The formation consists of two sandstone members, with an intervening shale member, and where all three outcrop, the sandstone components give rise to two cliffs, separated by a bench on the shale. In its southern development the upper sandstone is made up of massive, white-to-cream, crossbedded sandstone. The few concretions present are thin and horizontally extended. A discontinuous lignite seam is found near the top at some localities. The surface of the cliff walls weathers into an arabesque of hollowed and bossy sculpture. The thickness of this upper sandstone in the south is 130 feet, but due to replacement by shale above, it thins northward and near the mouth of Cadotte river is only 90 feet thick. At this locality the upper, massive, crossbedded, freshwater sandstone of the south is replaced by bedded sandstone and shale with marine fossils. Both the thickness and arenaceous content continue to decrease northward until the entire member is replaced by shale; so that the contact with the St. John shale descends stratigraphically in that direction.

" The middle shale member is made up of blue-black, friable shale without fossils, but probably is of marine origin. The thickness is 30 feet.

" The lower sandstone member differs considerably in structure and lithology from the upper. .At the top it is characteristically massive and crossbedded and contains large spherical concretions similar to those of the lower part of the Grand Rapids sandstone of the Athabaska section. This passes down into bedded sandstone and shale. The shale is thin-bedded at the base and carries marine fossils. The top may be subaerial, but the lower part is certainly marine. The contact with the Loon River shales is arbitrarily chosen, being marked by a gradual transition from bedded sandstone and shale to shale below. South of Brown's trading post the thickness is about 160 feet, 7 miles below the mouth of Battle river it is 80 feet thick, and to the north it is only about 20 feet thick. Both this formation and the Loon River have a common fauna and this is now being studied. The affinities of this fauna are Lower Cretaceous (Albian and Aptian rather than Neocomian) and pre-Dakota, taking Cenomanian as the base of the Upper Cretaceous. As to Dakota there are three possibilities: it is represented by the top few feet of the Peace River sandstone; the lowest beds of the St. John are its marine equivalent; or there was no deposition in Dakota time.

. " St. John Formation. The St. John shales form the gentle valley slopes above the cliffs made by the Peace River sandstones and underlie the adjacent plateaus, from the great horseshoe bend in the north to some distance south of the town of Peace River. There the Dunvegan sandstone comes in and underlies the plateau for some distance to the south. The St. John shale, however, continues to outcrop along the valley sides southwestward to the bend near the mouth of Burnt river, but occupies lower and lower elevations above river-level in that direction. It consists largely of dark blue to grey, friable shale with occasional rounded or banded ironstone concretions and is unfossiliferous as far as known. At Peace River the formation is estimated to have a thickness of from 500 to 540 feet, but a more exact figure cannot be given until the section on Smoky river is studied. The Dunvegan sandstone and Smoky River shales overlie the St. John shales in the direction of Dunvegan. This formation is correlated with the Colorado group of Upper Cretaceous time.

"*Triassic*. Owing to lack of time it was not possible to work out the details of the Triassic stratigraphy. A reconnaissance,. however, revealed some new facts that are worthy of record. Dark purple limestones, hardened sandstones, and shales with *Pseudomonotis subcircularis* Gabb, outcrop at Rapide-qui-ne-parle-pas and also on the south bank of Peace river a couple of miles below the mouth of Ottertail river. About 4 miles above Fish creek on the north bank the strata exposed are steeply inclined to the west. Above are brownish weathering sandstones with stem and tree-trunk impressions. These are underlain by *Pseudomonotis*-bearing beds and probably mark the contact between the Triassic and the base of the Bull Head Mountain sandstone. At the mouth of Twentymile creek a second and probably lower Triassic horizon outcrops. This consists below of dark limestones, sandstones, and shale with *Dawsonites* and other ammonites. A little higher are lighter limestones and sandstones with *Terebratula*, etc. Previously the *Dawsonites* zone had not been recognized south of Liard river. The entire Triassic series must be quite thick and is probably all of marine origin.

"*Bull Head Mountain Sandstone*. The Bull Head Mountain formation consists of a thick series of strata of freshwater origin lying between the Triassic shale below and the St. John shale above. It appears first in the canyon midway between Deep and Johnson creeks and continues to the west as far as Twentymile creek. As a result of the preliminary examination, this series of rocks may be divided into two members. The upper member consists of sandstones, shales, and coal beds and is well exposed in the canyon and on Gething and Johnson creeks. The lower part is made up of massive, coarse, crossbedded sandstones and is exposed at the head of the canyon, on Portage and Bull Head mountains, and the high hills to the west. The sandstones overlying the *Pseudomonotis* beds about 4 miles above Fish creek probably represent the base of this formation. No fossils have been found in the lower part, but a few plants were collected in the upper coal-bearing shales and sandstones. These include a few cycads, conifers, etc., and a single specimen of a dicotyledon. Though it is not possible now to make any definite correlation with southern Alberta, the plant association of this flora suggests that of the lower part of the Blairmore formation of the Crowsnest district. The rich angiosperm flora (Dakota), however, that characterizes the top of the Blairmore is not found here. The Bull Head Mountain formation should be correlated with the Peace River, and probably also the Loon River formations of the eastern succession, since it occupies the same stratigraphic position; it is, therefore, of Lower Cretaceous age.

"*St. John Formation*. The St. John formation, as here interpreted, embraces all the strata lying between the Bull Head Mountain below and Dunvegan sandstones above. It consists of two thick shale members separated by a thin sandstone member. The lower shale unit consists of 800 feet of dark, thin-bedded, slightly arenaceous shale, and first appears in the canyon about midway between Johnson and Deep creeks. It is well exposed on the Hudson Hope anticline in the cliffs at Hudson Hope and in the lower part of Maurice creek. Below the Gates the structure carries it below river-level. The thin, even bedding points to marine conditions, although no fossils have so far been found.

"The middle sandstone consists of 50 to 80 feet of massive, crossbedded sandstone with vertical rootlets and prostrate stems and evidently repre sents temporary subaerial delta conditions. It forms steep-sided island. and cliffs at the Gates and above Hudson Hope, and outcrops in the Maurice Creek gorge.

"The upper shales are first seen at Deep creek. To the Gates they occupy the valley sides above the middle sandstone. From there to Cache creek and from near King's ranch to below North Pine river they underlie the entire valley slope and from Cache creek to below King's ranch and from near the mouth of Kiskatinaw river to below Montagneuse river they occupy the part of the valley side below the Dunvegan sandstone. They also underlie the plateau north of the river from the foothills to the Cache Creek escarpment and the plateau south of the river as far east as the South Pine at least. From the Gates to a little below the mouth of the Kiskatinaw river a complete section of the upper shale is obtained owing to the east dip. Below the gates it is made up of thin-bedded, dark, hard, and friable shale with banded concretions of iron-stone. Fossil-bearing, dark, friable shales with concretions are found near river-level from a few miles below the mouth of Cache creek to within about 4 miles of North Pine river. Overlying this and appearing within about 2 miles of the river are black, paper-thin shales without fossils. Above this again come dark, thin-bedded, arenaceous shales with several large sandstone lenses. Downstream and higher are thin-banded sandstone and shale to the contact with the Dunvegan sandstone. Farther downstream there is much lateral replacement of sandstone and shale along this contact. The thin, even bedding and marine fossils point to marine conditions of deposition. The upper shale member is estimated to have a thickness of 1,300 feet in the vicinity of Cache creek.

"The small fauna of this formation is correlated with that of the Colorado group of the Upper Cretaceous.

"*Dunvegan Formation.* The Dunvegan sandstone first appears in the escarpment at Cache creek, but a few miles downstream the higher elevations upholding this formation recede toward Charlie lake. It reappears in high hills in the north bank a little below the mouth of Kiskatinaw river. Rapidly descending on the valley sides, it forms cliffs to within a short distance of river-level. Below Montagneuse river and continuing past Dunvegan it outcrops to river-level. The formation is made up of light, massive, crossbedded, soft sandstones with large flat concretions, and weathers into castellated forms. The crossbedding is on a scale of about 2 feet, truncated above, tangential below. Grains consist of quartz, feldspar, a little mica, and a black mineral. Other areas of this formation are more argillaceous and locally thin-bedded. A thin lignite seam in the sandstone varies in thickness, but never exceeds 6 inches. There is also a rather prominent bed of shelly limestone, chiefly of freshwater shells, in the cliffs below Montagneuse river and below Dunvegan. The crossbedded structure, coal beds, and freshwater fossils point largely to subaerial conditions of deposition, but the presence of *Inoceramus* and *Ostrea* indicates partial brackish water conditions, and probably the sea was not far away at any time. The formation has a thickness of about 530 feet. Both above and below it exhibits gradational contacts with continuous formations.

" This fauna includes *Ostrœa anomioides, Barbatia micronema, Brachydontes multilinigera*, and *Corbula pyriformis*. It is correlated with the Colorado group of the Upper Cretaceous.

Lateral Changes in Sedimentation.

" From Peace river northward three changes are noted: a decrease in thickness, replacement of sand by shale, and the substitution of subaerial by marine conditions of deposition. In the section from the mountains eastward the most striking change is the decrease in thickness of the St. John shales. The horizon of the subaerial sandstone member of the St. John in the west is probably replaced by marine shale in the eastern part of the section. In the west the entirely subaerial Bull Head Mountain sandstone appears to be the shoreward equivalent of the Peace River and Loon River formations."

The following paragraph by G. A. Dawson[1], descriptive of the rocks of Pine River South and Table mountain, is of interest.

" In the Pine River cañon, the rocks of this subdivision are flaggy sandstones, often brownish-grey in colour and false-bedded or ripple-marked, greenish-grey, fine-grained sandstones and black, soft, argillaceous sandstones and shales holding plant impressions, also occur. In the valley of a small stream which cuts the bank on the south side of the cañon, not far above the river level, Mr. Selwyn, in 1875, found, in alternating strata of sandstones and shales, four thin seams of coal, which in descending order are—six inches, eight inches, two feet, and eight inches thick. A number of fossils were also found in the associated beds, consisting of leaf-impressions and shells. 'The former occur chiefly in beds below the coal seams, and the latter in the intervening sandy shales, and in the ferruginous and calcareous concretionary nodules which accompany the latter.' These coal seams and the associated beds are at least 1,700 feet below the sandstones of the summit of Table mountain, and as the beds are nearly horizontal, this difference in elevation must closely correspond with the actual thickness of the rocks. For a portion of the ascent of Table mountain, however, the rocks are not seen, though about 200 feet thick of sandstone caps the hill. It is, therefore, uncertain whether the subdivision classed as the Upper Shales[2] may occur in the concealed interval and the sandstones at the summit represent the Upper Sandstone[3] series, or whether—as is perhaps more probable—the entire thickness of the rocks from the edge of Pine river to the summit of Table mountain, should be classed as belonging to the Lower Sandstones[4], which in this case must here have a very great thickness. In the sandstones of the summit of the mountain, numerous specimens of *Inoceramus altus*, a species previously noted, in beds supposed to represent the Pierre group in Wyoming, were found by Mr. Selwyn."

The Smoky River shales, or Upper Shales of Dawson, overlie the Dunvegan sandstone and have a width of about 36 miles on Smoky river and an approximate thickness of 350 feet. These shales, with a thickness of 200 feet, are also exposed on Coal brook 5 miles east of the lower

[1] Geol. Surv., Can.; Rept. of Prog., 1879-80, p. 117B.
[2] Smoky River.
[3] Wapiti River.
[4] Dunvegan.

forks of Pine River South. The formation consists of greyish and bluis to nearly black shales holding ironstone in abundance. The fossils ar typical Pierre and the formation corresponds to the upper portion o the La Biche shales of Athabaska river.

The Wapiti River sandstones consist of soft sandstone, shaly sand stone, shale, and clay. The sandstones are often nodular and hold band of ironstone, coaly fragments, and obscure fossil plants. They are yellow ish, greyish, or bluish-grey in colour, and the shales and clays have ofte a brownish earthy appearance. This formation overlies the Smok, River shales and is exposed on Smoky river and Wapiti river[1]. It ha. a thickness of approximately 200 to 300 feet and may correspond to the Edmonton farther south. Dowling, however, suggests a different correla tion for this and other formations on the Peace.[2]

McConnell gives the following description of the Cretaceous formation of the Liard and Mackenzie rivers.[3]

" Fossiliferous Cretaceous beds were not recognized in descendin the Liard until the plateau belt which borders the eastern foot-hills wa reached. Below Fort Halkett, west of the mountains, a band of soft dark shales crosses the river, which may be in part Cretaceous, but n fossils were found in .it. The eastern foot-hills are built of a great serie of alternating shales and sandstones, with some limestones, all folde closely together, which resemble those found in the foot-hills farther south and, like them, probably consist largely of Cretaceous rocks, but it was found impossible on a hasty trip along one line to separate these from the Triassic, or from the shales which cap the Palæozoic system, owing to the lithological similarity which prevails throughout. East of the foot-hills the convolutions gradually cease and the section becomes more legible. The beds here consist of soft, finely laminated shales, interstratified with a few beds of sandstone and ironstone. They have a minimum thick- ness of 1,500 feet. The shales yielded some fossils among which were several specimens of *Placenticeras Perezianum*, one of the characteristic fossils of series C. of the Queen Charlotte islands. With this were species of *Camptonectes* and *Inocerami*. Near the eastern edge of the plateau belt the shales are overlain by massive beds of rather soft sandstones and conglomerates, the thickness of which was not ascertained. The conglom- erates are affected by a gentle easterly dip, and descend to the level of the river in the course of a few miles. From the point at which they disappear to the eastern edge of the Cretaceous basin, the rocks consist of dark fissile shales, crumbly sandy shales, and sandstones, but the exposures along the valley are infrequent, and the succession soon becomes obscure.

" The Cretaceous section along the Liard thus shows two great shale and sandstone series separated by a heavy band of sandstones and con- glomerates. The lower shales, from the imperfect fossil evidence at hand, and also from their lithological character, may be referred tenta- tively to the horizon of the Queen Charlotte Islands or Kootanie form- ation, the upper shales to that of the Benton, while the intervening con- glomeratic band probably represents the Dakota. The lithological suc- cession of the Cretaceous beds here is almost identical with that which

[1] Dawson, G. M., Geol. Surv., Can., Rept. of Prog., 1879-80, pp. 116B-125B.
[2] Roy. Soc. Can., Trans., ser. 3, vol. IX, sec. 4, pp. 27-42.
[3] Geol. Surv., Can., Ann. Rept., vol. IV, pp. 19D-21D.

obtains in other parts of the Cordilleran belt north of the International Boundary and on the Queen Charlotte islands[1] and shows that similar conditions of deposition prevailed at the same time over this whole area.

"The Cretaceous rocks cross the Liard with a width of over a hundred miles, and north of the river enter a bay in the mountains, the extent of which to the northwestward is not known; southwards they are connected with the great Cretaceous basin of the plains.

"Fifteen miles below Fort Liard the Devonian limestones rise to the surface, but the junction between them and the overlying Mesozoic rocks is concealed, and I was unable to ascertain whether the older beds continue to the eastern edge of the basin or are here overlapped by the Upper Cretaceous. It is probable, however, that the latter is the case.

"The plains bordering the lower part of the Liard and the upper part of the Mackenzie rest on Devonian limestones and shales, and Cretaceous rocks were not detected in descending the latter stream until the Dahadinni river in latitude 64 degrees north was reached. They consist here of a couple of hundred feet of dark grey shales and sandstones. They are exposed along the valley for ten or twelve miles, and are then concealed by the boulder-clay, but probably continue under the latter as far as the Tertiary basin at the mouth of Bear river, a distance of fifty miles. The Cretaceous beds here occupy a depression between two high ranges of limestone mountains and cannot have a greater width than ten or fifteen miles. They have been separated from the Cretaceous beds which form the western shores of Great Bear lake by the elevation of the Mount Clark range.

"Forty miles below Bear river the Cretaceous beds reappear on the banks of the Mackenzie, and with the exception of one break of a couple of miles where they have been removed by denudation, underlie the valley all the way to the Ramparts, a distance of ninety miles. The fossils obtained from this area and from the one above Bear river consist of fragments of Ammonites and Inocerami, too imperfect for specific determination.

"A hundred and twenty miles below the Ramparts, the Mackenzie enters a third Cretaceous area, and the largest one on the river. Cretaceous beds appear in the banks a short distance below old Fort Good Hope and extend down the Mackenzie to the head of its delta, and westwards across the Rocky mountains and down the Porcupine to about longitude 139 degrees west. They consist on the Mackenzie of coarse shales interstratified with some sandstones and fine-grained conglomerates; in the mountains of several thousand feet of barren sandstones and quartzites underlain by dark shales, and on the Porcupine of the same two series underlain by a great thickness of alternating shales, sandstones, and conglomerates, holding *Aucella Mosquensis* var. *concentrica*. The intermediate dark shales are probably of Benton age, while the lower division so far as the fossil evidence goes, represents the Queen Charlotte Island formation and the Dakota."

Coarse-grained sandstones are exposed along Fort Nelson river for 12 or 14 miles up from a point 12 miles above its mouth. Black shales are seen 20 miles above Fort Nelson; farther upstream these are overlain by sandstone, and at a point 60 miles above Fort Nelson they disappear at the

---

[1] Am. Jour. Sc., vol. XXXVIII, p. 120.

water's edge beneath the sandstone. On Sikanni Chief river, 45 miles above its junction with the Fort Nelson, the banks are composed of shale and sandstone. Farther up, the river has cut a canyon, which at a point near the headwaters of Pine river is 1,100 feet deep. The lower 475 feet of strata here exposed consist of black and grey shale with some thin layers and masses of clay ironstone. This is succeeded upward by sandstone. Cliffs of this sandstone occur on Pine river and other streams between Sikanni Chief and Peace rivers.[1]

Some soft sandstones and conglomerates, probably of Cretaceous age, are exposed for a few miles along the north bank of Gravel river about 25 miles from the Mackenzie.[2]

The greater portion of the basin of Peel river below Wind river is underlaid by soft Cretaceous shale and sandstone. Between the lower canyon and Snake river the Peel has cut a valley 500 to 700 feet deep in these sediments, and about 5 miles below the canyon a section in the bank shows 200 feet of yellow and red shales, which towards the base are interbedded with layers of sandstone. This is succeeded downward by massive sandstone 50 feet thick, underneath which occur about 150 feet of very fissile, rusty, pyritous shale. The glacial drift in this section is about 40 feet thick. Northward the shales gradually increase in thickness until they predominate over the sandstones. The sediments lie in gentle undulations. A good section of Cretaceous rocks is obtained in mount Goodenough, which is 3,000 feet high and lies to the west of the Mackenzie delta, 2 miles from Huskie river, the western branch of the Peel. The strata are horizontal or only slightly folded. "At the base is a thick series of black shales, which towards the top contain beds of very hard clay ironstone. These weather red, and the outcrop can be traced by its colour for miles along the eastern face of the mountains. These red beds contain remains of Ammonites, while the underlying and enclosing black shales are also fossiliferous. The shales are gradually replaced upwards by argillaceous sandstones, and these again by siliceous sandstones. These latter become metamorphosed to quartzites and constitute the upper members of the series."[3]

Nearly horizontal Cretaceous sediments surround the western and northern portions of Great Bear lake, reaching to within 30 miles of Fort Confidence, and extending down Great Bear river to within 7 or 8 miles of its mouth, being replaced, however, at the rapid by Palæozoic strata that form the range of mountains cut by the river. Around the lake there are few outcrops of rocks of this formation. Shales and sandstones are exposed along Smith bay, and the Sweet Grass hills represent a low anticline, composed of hard sandstone, which forms Gros Cap peninsula. Above the rapid on Great Bear river there are numerous exposures of rocks in which some fossils are found.[4] The western part of the peninsula north of McTavish bay is probably underlain by Cretaceous sediments, and Richardson's notes indicate that Cretaceous sediments are exposed in Grizzly Bear mountain which forms the promontory between McVicar and Keith bays.

[1] Ogilvie, Wm., Dept. of Interior, Ann. Rept., 1892, pt. 7.
[2] Keele, Joseph, "A reconnaissance across the Mackenzie mountains on the Pelly, Ross, and Gravel rivers, Yukon and Northwest Territories," p. 40. Geol. Surv., Can., 1910.
[3] Camsell, Charles, Geol. Surv., Can., Ann. Rept., vol. XVI, p. 46CC.
[4] Bell, J. M., Geol. Surv., Can., Ann. Rept., vol. XII, p. 25C.

Tertiary.

Tertiary sediments occupy a few small areas in the Mackenzie basin.

The Paskapoo formation is exposed on Pembina river near longitude 115 degrees west, where a thickness of 50 feet or more of yellowish-grey sandstones overlies 70 feet of concealed rock resting upon a seam of coal. The formation extends some distance to the west.[1]

"Tertiary beds occur at the mouth of Bear river and occupy a basin of about thirty to forty miles in length and twenty to thirty in breadth. They rest unconformably on the underlying Cretaceous shales and Devonian limestones. They are lacustral in origin and consist largely of discordantly bedded sand, sandy clays, clays, and gravels. Beds of purely argillaceous material, usually somewhat plastic in character, are also present, and seams of lignite and carbonaceous shales not infrequently constitute a considerable portion of the section............The beds have an anticlinal attitude on the whole, but are usually nearly horizontal. They have a minimum thickness of 600 feet."[2]

A small area of Tertiary sediments occurs on Peel river in the vicinity of Wind and Bonnet Plume rivers and reaches from 20 to 30 miles up the valleys of these rivers. The rocks consist of sandstones and clays carrying beds of lignite. These have an unconformable relation to the underlying strata, and have been gently folded into a number of anticlines and synclines.[3]

Pleistocene and Recent.

With the exception of the higher points of the Rocky and Mackenzie mountains on its western side, the whole basin of Mackenzie river was overridden by glacial ice during a part at least of the Pleistocene period. Ice appears to have entered the basin from two sides, namely, from the Keewatin centre on the east which is believed to have occupied a position on the northwestern side of Hudson bay, and from the Cordilleran ice-sheet on the west, the centre of which lay to the west of the Rocky mountains in the northern part of the province of British Columbia. At the same time tongues of ice flowed down the valleys on the eastern side of Mackenzie mountains, rising from a confluent ice-sheet which lay on the western slope of those mountains and formed the northern extension of the Cordilleran ice-sheet.[4] Keele states that "the valley of the Mackenzie river was occupied by an ice-sheet of considerable thickness which pushed up the valley of the Gravel river before the ice from the Cordilleran glacier began to pour down."

Ice from the western sources entered the basin of Mackenzie river in the form of valley glaciers which flowed down the valleys now occupied by the western tributaries of the river. Keele has estimated that a thickness of 2,000 feet of ice occupied the valley of Gravel river in its lower part, and on Wind river, a tributary of the Peel, Camsell found that the maximum thickness of the ice in that valley must have been about 1,000 feet. Ice tongues also occupied the valleys of the Athabaska, Pine, Peace, and Liard

[1] McEvoy, James, Geol. Surv., Can., Ann. Rept., vol. XI, p. 24D.
[2] McConnell, R. G., Geol. Surv., Can., Ann. Rept., vol. IV, p. 22D.
[3] Camsell, Charles, Geol. Surv., Can., Ann. Rept., vol. XVI, p. 41CC.
[4] Keele, J., "A reconnaissance across the Mackenzie mountains", Geol. Surv., Can., 1910.

MICROCOPY RESOLUTION TEST CHART

(ANSI and ISO TEST CHART No. 2)

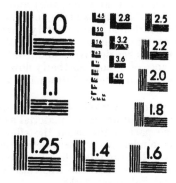

APPLIED IMAGE Inc

1653 East Main Street
Rochester, New York 14609   USA
(716) 482 - 0300 - Phone
(716) 288 - 5989 - Fax

rivers in the mountains and foothills portions of their courses and probably spread out as piedmont glaciers on entering the more level country of the plains.

On the eastern side, ice from the Keewatin centre invaded the Mackenzie basin as a continental ice sheet, its encroachments gradually extending farther and farther westward until it joined with the ice from the western sources. From the evidence of striations and transported material the direction of movement of the Keewatin ice sheet is shown to have been southwestward in the southern part of the basin and westward and even northwestward in the northern part. South of Athabaska lake, for example, the general course of the ice was south-southwest; immediately north ot Athabaska lake it was southwest; and on Great Slave lake it was almost west. Farther north, in the valley of Mackenzie river, about latitude 66 degrees north, McConnell[1] noted evidence of a more northerly flow of an ice sheet approximately 1,500 feet thick, which extended down the valley of the river and thence out to sea.

The results of glacial action in the basin of Mackenzie river are everywhere shown by the presence of various kinds of deposited material as well as by the eroded surfaces. Over a large part of the basin the effects of erosion are predominant, and over the remainder of the basin deposition of eroded and transported material prevails. The boundary line between the region of erosion and that of deposition follows more or less closely the boundary between the Pre-Cambrian and Palæozoic rocks.

The region of erosion is characterized by bare rock surfaces, rounded and glaciated by the passage of the ice sheet over them. Deposited material is found in the hollows of the surface and consists mainly of boulder clay. Moraines, drumlins, eskers, sand-plains, and some curious drift hills called "ispatinows"[2] by J. B. Tyrrell occur here and there, but are not notably abundant.

In the central and western portions of the basin, where deposition is more evident than erosion, the surface is in general so heavily drift-covered that the older, underlying rocks are nearly everywhere concealed. The glacial deposits have not been uniformly distributed but are thickest in old, preglacial depressions, whereas on the ridges they become greatly attenuated. Inequalities have been reduced and a rolling topography developed.

Sections of the glacial deposists of the central portion of the basin show in many cases boulder clays usually underlaid by stratified sands and gravels and frequently overlaid by a second series of sands and gravels. Such sections occur frequently in the basins of Athabaska and Peace rivers and in the valley of the Mackenzie. The stratified sands and gravels, both below and above the boulder clay, are, according to McConnell,[3] evidently lacustrine in origin and were doubtless deposited in greatly expanded lakes near the ice front, both before the advance and immediately after the retreat of the ice sheet, whereas the boulder clay represents the ground moraine of the ice sheet itself. Thicknesses of these beds vary considerably. Sections of the boulder clay occasionally show a thickness of 250 feet and the sands and gravels over and beneath the boulder clay are in some cases fully 150 feet thick.[4] Moraines are apparently fairly prevalent in this portion

[1] Geol. Surv., Can., Ann. Rept., vol. IV, 1888-89, p. 27D.
[2] Geol. Surv., Can., Ann. Rept., vol. VIII, 1895, p. 23D.
[3] Geol. Surv., Can., Ann. Rept., vol. V, 1889-90-91, p. 61D.
[4] McConnell, R. G., Geol. Surv., Can., Ann. Rept., vol. IV, 1888-89, p. 26D.

of the basin. McConnell,[1] describes a prominent morainic ridge southwest of the west end of Great Slave lake as consisting of a "medley of steep-sided, interlacing hills and ridges, similar in appearance to those found on the Grand Côteau de Missouri of the plains" and evidently of like origin.

On the final retreat of the ice sheet from the basin of Mackenzie river, lakes occupied not only the present lake basins but other large tracts of country which have since been filled up or drained. Filling of some of these lakes by the streams flowing into them is still in progress. Both the Peace and Athabaska rivers in the flood season carry a great deal of sediment and this is being deposited as fine sand, clay, or slit in their deltas, gradually filling up the western end of Athabaska lake. A former southern arm of Great Slave, which at one time extended up the valley of Slave river as far as Fort Smith, has already been filled with the same sort of material from Slave river. The Mackenzie itself carries some sediment in flood time and is building a great delta, the emerged portion of which is about 80 miles across its seaward side and 100 miles deep. The submerged portion extends in a broadening front far out into the Arctic ocean.

The rivers of the Mackenzie basin are very largely post-Glacial and have cut deep valleys into the Glacial and underlying Cretaceous shales and sandstones. In certain instances, as in the case of Little Buffalo, Hay, and Beaver rivers, gorges several miles in length have been cut into Devonian limestones and shales to depths of 100 or 200 feet, and as in the case of Niagara falls and gorge, give a scale by which estimates can be made as to the time that has elapsed since the recession of the ice-sheet.

Section in Peace River Pass.

In a hurried reconnaissance through Peace River pass it was found that the rocks exposed along the pass consist principally of "greyish Palæozoic limestones striking in a northwesterly direction, and dipping persistently to the southwest. Repetitions of parts of the limestone series, caused by overthurst faults, occur at several points." No infolds of Cretaceous strata, such as occur farther south, were observed.

[2]"Immediately east of the main range, exposures of yellowish-weathering, calcareous sandstone, probably of Cretaceous age, occur in the banks of the river. These are replaced going westward by greyish limestones dipping steeply to the west. The junction between the limestone and sandstone is concealed in the valley, but there is little doubt from the relative position of the two formations, that the contact is a faulted one and that the Palæozoic limestones of the mountains, here as elsewhere along the eastern boundary of the range, are thrust up over the Mesozoic rocks of the foot-hills.

"The limestones are fossiliferous, the fauna, so far as ascertained, being similar to the Banff or Devono-Carboniferous division of the Bow River section.

"West of the fault, the limestones stand at a steep angle, the beds being fairly regular, but farther west they become greatly confused and show evidence of much disturbance. In the second range, the limestones are overlain by a band of dark Monotis-bearing calcareous shales and impure limestones of Triassic age. West of the Triassic band, a second fault

[1] Op. cit., p. 25D.
[2] McConnell, R. G., Geol. Surv., Can., Ann. Rept., vol. VII, p. 32C.

brings the Banff limestones again to the surface, and the same limestones probably repeated by faults, occur in the next two ranges. In the first of these, the Banff limestones and overlying Triassic beds have a regular westerly dip, but in the second, a line of strong disturbance is reached, and the strata as seen on the mountain sides are crushed into numerous subordinate folds.

"A fault of considerable magnitude crosses the valley west of the two ridges referred to above, and brings up limestones which were referred to the Castle Mountain group. West of this fault, the dips north of the river for some distance were too confused to follow, but south of the river, the beds, with the exception of one double fold, dip regularly westward until near mount Selwyn. The limestones in this part of the range are mostly unfossiliferous and of the Castle Mountain type, but higher beds holding *Halysites* were found in one place.

"Mount Selwyn shows a sharp anticline on its eastern slope. The centre of the mountain is formed of almost vertical limestone beds, but going westward these are soon replaced by the quartzites, schists, and crushed conglomerates of the Bow River series. The latter are forced up over the limestones by a well-defined overthrust fault, running in a northwesterly direction.

"Mount Selwyn is flanked on the west by a small range composed partly of the rocks of the Bow River series and partly of the schists of the still older Shuswap series, all dipping to the southwest. The latter overlie the former, but the cause of their superior position was not ascertained."

Section on Liard River.

The following section along Liard river, from Dease river to the Mackenzie, is taken from McConnell's report of his traverse made in 1887.[1]

Three miles below the mouth of Dease river there is an exposure of soft, dark shales associated with friable shales and conglomerates probably of Tertiary age, and on a small island at the mouth of Hyland river there is an exposure of hard, whitish sandstone passing into quartzite and dipping northeast at an angle of 50 degrees. Six miles farther down, unconsolidated sands, sandy clays, and gravels holding small beds of impure lignite are exposed in a cut bank.

"The rocks in the Little Cañon consist of dark and sometimes cleaved shales, holding large, flattened ironstone nodules, hard sandstones and quartzites, and some beds of fine-grained, hard, siliceous conglomerate. They are closely folded together and strike north 35 degrees west. No fossils were found in any of these beds nor any definite evidence of their age obtained beyond the fact that they have a close lithological resemblance, both in appearance and composition, to those on Dease river, from which Dr. Dawson obtained graptolites of Utica-Trenton age."

A few miles below Little canyon the shales, sandstones, and conglomerates are replaced by shaly limestone. This is succeeded lower down the river by more massive varieties of limestone. Still lower down a range of hills extends in an irregular manner for some miles along the left bank of the river and rises to an altitude of 1,500 feet above the river, or 1,000 feet

[1] Geol. Surv., Can., Ann. Rept., vol. IV, pp. 33D–58D.

above the general plateau level. "The limestone of which the hills are formed is usually greyish in colour and rather compact, but passes in many places into a whitish, highly crystalline variety without distinct bedding. It has a general strike of north 15 degrees west. It is destitute of determinable fossils, but holds fragments of crinoid stems, and traces of brachiopods and trilobites."

Limestone outcrops are seen along the river to within a few miles of the mouth of Turnagain river. These limestones are often coarsely crystalline and cut by white calcite veins. "Other varieties show wavy lines projecting from weathered surfaces due to alternating magnesian and calcarcous layers, and closely resemble in this respect the limestones of the Castle Mountain group as developed along the Bow River pass. In some places the limestone becomes shaly and impure, and is altered into an imperfectly developed schist."

The limestone is replaced several miles above Turnagain river by shales, sandstones, and conglomerates, that are exposed at many points as far as Whirlpool canyon, above Coal river. These rocks resemble those of Little canyon and are evidently of the same age. The shales are dark and finely laminated and are interstratified in places with the sandstones. The latter are lighter coloured than the former, are hard, and often pass into quartzite. The conglomerates are very fine-grained and consist principally of white, well rounded, quartz pebbles, embedded in a siliceous matrix. Resting unconformably upon these beds at Cranberry rapids, a few miles above the mouth of Turnagain river, there is an exposure of soft shales and conglomerates, evidently of Tertiary age.

Lignite of inferior quality is brought down by Coal river, but although an examination was made of several miles of the bank its source was not discovered. The banks of the lower part of coal river are low and consist of uncemented sands, clays, and gravel similar to the lignite-bearing beds above Little canyon. This formation is of irregular thickness, but of wide distribution, and fills depressions in the older rocks from the mouth of the Dease to the passage of the Rockies.

"The rocks observed along the river from Whirlpool canyon to the lower end of Portage brûlé consist altogether of different varieties of limestone. This occurs in some places in massive beds, ranging in texture from compact to moderately crystalline. In other places it becomes very impure and shaly, and often passes into imperfectly developed calc-schists. No fossils were obtained from it, but it has a close lithological resemblance to the limestone occurring above Cranberry rapids."

Dark shales, softer than those farther up the river and probably of Mesozoic age, are exposed near the mouth of Smith river and can be traced several miles down the Liard, where they are finally replaced by limestone. The main range of the Rocky mountains where cut by the river consists of a greyish and moderately compact limestone. The strata first met are horizontal, but a short distance farther down they assume almost vertical attitudes and apparently are folded in a sharp anticline. The limestones are exposed for 6 miles along the river and then become concealed by dark shales similar to those farther down from which Triassic fossils were obtained. Shales of Mesozoic age are thus found to occur on both flanks of the mountain.

In the foothills, which here have a width of 38 miles, the rocks exposed consist chiefly of shales with bands of sandstone, and occasionally some

86

limestone. The shales undulate at all angles, and at the east end of Devil's portage an anticline brings to the surface greyish, moderately crystalline limestone holding fragments of crinoids and other fossil These limestones probably belong to the upper part of the Palæozoic series of the mountains. The shales are dark in colour, and as a rule are rather coarsely laminated. Fossils were collected at the "Rapids of the Drowned", at Hell Gate, and at the lower end of the canyon above Hell Gate, and were referred by Whiteaves to the Triassic. "The localities from which Triassic fossils were obtained, extend along the river for ten miles, but it is highly probable that the rocks of this age have a much wider distribution than this and include the greater part of the barren shales above the 'Rapids of the Drowned,' as well as those below Hell Gate. It will require, however, more time than could be spared on a rapid reconnaissance, to separate precisely the shales which cap the Palæozoic from those of the Trias, and the latter from the Cretaceous."

The plateau district to the east of the foothills consists of flat-lying, Cretaceous shales. They are dark in colour, are soft and finely laminated, and are interstratified with small beds of sandstone and ironstone, and layers of ironstone nodules. Towards the eastern part of this plateau belt the shales along the river are overlain by massive beds of soft sandstone and conglomerate, which form a steep escarpment running parallel with the river.

The river leaves the plateau by a narrow gap a short distance above the mouth of Beaver river and enters much lower country, where occasional exposures occur of Cretaceous shale and sandstone and of Devonian shale and limestone.

Section on Gravel River.

The following description of the rocks exposed on Gravel river is taken from Joseph Keele's "Reconnaissance across the Mackenzie mountains on the Pelly, Ross, and Gravel rivers, Yukon, and Northwest Territories.[1]

"The mountains of the watershed at the head of the Ross river are formed of alternating beds of dark, compact quartzite and grey shale and slate.

"About ten miles east of the divide some yellowish, crystalline limestones occur in low isolated cliffs along the embryo Gravel river, but the principal rocks are dark, sandy shales, striped grey slates, and micaceous sandstone or quartzose schists. Rocks of this character extend eastward to Mount Sekwi, about fifty miles from the divide, and then end abruptly....

"A radically different geological province begins at Mount Sekwi, and limestones, dolomites, sandstones, and conglomerates, etc., of various bright colours, replace the sombre rocks to the westward."

The eastern part of mount Sekwi is composed of dove-coloured limestones. These pass into grey argillites and are interbedded with sandstone or quartzite. Farther east there is a section of over 2,000 feet of rather pure limestone. These rocks are of Silurian age and vary in attitude from nearly horizontal to vertical.

Purple and greenish argillites, followed upward by dolomite, calcareous sandstone, and limestone occur at the junction of the Natla and Gravel rivers. These beds, which are of Cambrian age, have a total thickness of

[1] Geol. Surv., Can., Pub. No. 1097.

4,000 feet and are inclined at a low angle to the west. Dolomite forms a large part of this section.

Below the mouth of the Natla the mountains are composed of rocks quite different from the foregoing, beneath which they appear to dip. They are probably of Cambrian age and consist of

	Feet.
Brown, micaceous, sandy slates	1,100
Conglomerate	2,000
Coarsely laminated hematite and siliceous slate	100
Dolomite and argillite	1,000

The conglomerates form the wall rock of Shezal canyon.

Below Twitya river the rocks composing the Tigonankweine range consist of 4,000 feet of alternating beds of argillite, dolomite, and limestone, succeeded upward by a sill of diabase, 100 feet thick, and 1,500 feet of sandstone. The diabase sill has a horizontal extent of several miles. This series, which is referred to the Ordovician system, lies nearly horizontal. The sandstones, whose prevailing colour is reddish, grow thicker towards the east and opposite the mouth of Nainlin brook are about 4,500 feet thick, forming the entire mountain mass. They extend down the river as far as Inlin brook where they are replaced by limestones.

These limestones are probably of Devonian age. They are more or less massive, are broken into several fault blocks, dip to the southwest, have low escarpments facing northeast, and constitute the eastern foothills of Mackenzie mountains. With decrease in the altitude of the foothills to the eastward the limestones become less tilted and broken, the bedding is thinner, and shaly layers appear. In some of these, fossils of Devonian age were found.

"About twenty-five miles from the Mackenzie some soft sandstones and conglomerates are exposed for a few miles along the north bank of the Gravel river; the beds are inclined slightly toward the west, and have a thickness of about 200 feet.

"The sandstones are coarse-grained and nodular, of yellowish or grey colour, grading into fine conglomerates, which are made up chiefly of black cherty argillite fragments." They are probably of Cretaceous age.

Section on Peel and Wind Rivers.

In the section along Wind river, from Nash creek to the edge of the mountainous country, the following succession of rocks occurs in ascending order: ferruginous slates and argillites; limestones, and sandstones with some limestones; and reddish conglomerate.

At the mouth of Nash creek the rocks consist of closely folded black slates with remnants of the overlying limestone occurring in the synclines. They strike east and west across the valley of the river and dip at high angles or are vertical. Northward the limestone becomes gradually more dominant and the underlying slate appears only when brought up by an anticline. Fifteen miles below Nash creek, a coarse-grained, white sandstone is met overlying the limestone and slate. The strata are horizontal or dip at a low angle to the north. At this place the limestone is not so thick as farther up stream and appears to rest unconformably on the slate. From Bear river to the point where Wind river leaves the mountains, sandstone and limestone appear in a succession of gentle anticlines and syn-

clines. A small body of dark reddish conglomerate overlies these on th
edge of the slope.

On emerging from the mountains, Wind river enters a broad platea
The level of this plateau is broken in places by several short ranges of mou
tains, which are really the foothills of the main range. The foothills are
extends northward some distance beyond the east and west stretch of Pe
river and eastward to Snake river. Almost in the centre of this area li
a large basin of over 500 square miles, occupied by slightly disturbe
Tertiary rocks. This basin, which is almost completely enclosed by foo
hills, lies between Wind and Bonnet Plume rivers and extends southwan
from the Peel some 50 miles.

The geology of the foothills section is in marked contrast with the
of the mountain section. Three miles from the base of the main range the
are cliffs 150 feet high composed of slightly inclined beds of fine conglomera
holding fragments of fossil wood and ironstone nodules and merging in
soft grey sandstone. The latter becomes feldspathic towards the top, ar
is overlain by boulder clay. These sandstones, which are probably
Cretaceous age, form cliffs on both sides of the river down to within 2 mil
of Little Wind river.

The Illtyd range consists of massive, grey, dolomitic limestones havii
an anticlinal structure. "Mount Deslaurier is a west-facing fault scar
rising abruptly from the water's edge to a height of 850 feet. It is con
posed of about four hundred feet of dark reddish conglomerate, containii
angular and water-worn fragments or limestone, quartzite, and oth
rocks; below this is a brecciated limestone, which, near the contact, al
carries some foreign fragments. At the water's edge is some sandstone
Mount Deception, standing at the junction of Hungry creek and Wi
river, is composed of massive crystalline limestone. Below mount Dece
tion the river enters the level country underlain by almost undisturbed T
tiary clays and sandstones.

A seam of lignite 6 feet thick is exposed 12 and 14 miles below mou
Deception. Four miles above Pee' river a section is exposed showi
5 feet of rusty black slates ove: ... formably by 50 feet of san
stone, with eight seams of lignit ' nch to 4 inches thick. Ove
lying this there is a thickness of 4. ... vel and boulder clay. One m
above Peel river the contact betw ... iary sandstones and the und
lying black slates is well exposed, mer dipping at a low angle to t
east and the latter being vertical.

On Peel river, tilted, fine-grained slates extend from several mi
above to three-quarters of a mile below the mouth of Wind river, whe
they are replaced and overlaid by Tertiary clays and sandstones. T
width of the Tertiary basin where it is crossed by the Peel is 13 mil
The formation consists of clays and sandstones carrying beds of ligni
It is gently folded into a number of anticlines and synclines. One lignite b
near the top is 30 feet thick and fairly persistent, being exposed at two poin
4 miles apart with a shallow syncline between. It rises in an anticline, t
top of which has been eroded away, and beyond dips again and disappea
beneath the bed of Bonnet Plume river. It has been burnt for some d
tance along the bank of the river.

At the lower canyon the river cuts the same series of slates as a
exposed in the upper canyon. A band of crystalline limestone apparen

forming the base of the series is exposed at a point halfway through the canyon and at the lower end. No fossils were found in these rocks.

From the lower canyon to Snake river the Peel flows through a valley cut into soft shales and sandstones of Cretaceous age. Below Snake river argillaceous sandstones and clays are exposed, and below Trail creek soft shales, often pyritous, occupy a large section of the banks and are associated with a sandstone carrying fragments of Ammonites. The Peel leaves the plateau at Satah river and enters a low-lying, level country underlain by soft sandstones and some conglomerate. Only alluvial sands and clays are exposed below Fort McPherson.

Section along Peel River Portage West of Fort McPherson.

"The Rocky mountains, along the Peel River portage, present features which differ greatly from those which characterize them farther to the south. They consist here essentially of two ranges, separated by a wide longitudinal valley, and flanked on either side by high plateaus. The eastern range has a width of seven miles, and its higher peaks were estimated to reach an altitude of 2,500 feet above the level of the pass, or about 4,000 feet above the sea. The western range is much narrower, and north of the pass does not exceed four miles in width, but spreads out somewhat more towards the south. The valley of Peel river, which skirts the eastern base of the range, is fully 1,200 feet lower than the valley of Rat river on the western side, and the drainage of the mountains is mostly towards the former.

"The geological section obtained is somewhat imperfect, as our scanty supplies allowed of no delay, but sufficient was learnt to show that the range has on the whole an anticlinal structure, although the general anticline is obscured in places by subordinate folds, and is probably broken by faults. In the eastern plateau the beds are nearly horizontal, but approaching the mountains they incline greatly to the eastwards, and in the centre of the eastern range have dips of from 30 degrees to 70 degrees in the same direction. In the western range the same dip prevails, but the inclination is much less, and the beds flatten out when the mountains are replaced by the elevated western plateau. The horizontal altitude is maintained for some miles, but before reaching the western edge of the plateau the beds bend down and dip gently to the west.

"No limestones were observed along this section, and the rocks consist of sandstones, quartzites, and shales, all of which are probably referable to the Cretaceous. At the starting point on Peel river the banks are formed of shales, interbedded with some hard sandstones holding carbonized fragments of wood and leaves. In the first fourteen miles the beds are concealed, but shales and sandstones are again exposed on the banks of the valley in which we made our first camp. Some fossils were collected here, among which is a *Discina* like *D. pileolus*, Whiteaves, a *Mactra* and a *Yoldia*, both of which are probably new, but the specimens are too imperfectly preserved to admit of specific determination. Six miles farther west, at the entrance to the "Gap", the trail passes over beds of a bluish, rather compact calcareous sandstone. The beds are coated in places with calc-spar, are highly ferruginous, and weather to a rusty yellow. A *Cardium* and some other poorly preserved fossils were obtained

here. In the valley of the river the sandstones are underlain by dark
shales. After entering the mountains, only alternating sandstones and
quartzites were seen. The beds of this series are greyish in colour, are
evenly stratified, and are very uniform in appearance all across the range.
They have an estimated minimum thickness of 5,000 feet, and may possibly
greatly exceed this. The western plateau is built of compact, greyish
sandstones, passing in places into quartzites, similar to those found in
the mountains and evidently belonging to the same formation.'

STRUCTURAL GEOLOGY.

In the Pre-Cambrian rocks of the Laurentian Plateau portion of the
Mackenzie basin our knowledge of structure and structural relations is very
imperfect and is limited to the region about Athabaska lake and the territory along the Tazin and Taltson rivers. Beyond this, little mention is
made of structure in the reports of explorers.

Throughout the whole Pre-Cambrian portion of the basin, however,
the same structural relations seem to hold in a general way. Great granite
batholiths have been intruded into an older series of stratified rocks, of
which there are now only remnants left. These two series have together
suffered from severe orogenic disturbances, so that the stratified rocks are
now tilted at high angles and the granites have in part become gneissoid.
The remnants of the stratified rocks seem to be synclinal in structure and
the strike ranges from northeast on Athabaska lake to north and even
northwest on Taltson river. The trend of the gneiss conforms in a general
way to the strike of the stratified rocks, and in certain localities there is
much local contortion of the beds or foliation planes.

The younger rocks of the Pre-Cambrian, as illustrated by the Athabaska sandstone on the south shore of Athabaska lake and others on Great
Slave lake, show little evidence of disturbance. The beds are horizontal
or dip at very low angles. They are cut by later dykes which, however,
have created little disturbance and effected only slight local metamor-
phism. There is ssive jointing of the beds, but apparently little faulting.

In the great central plain there has been no igneous intrusion in the
rocks outcropping at the surface, and little orogenic disturbance, so that
the structure is the result mainly of simple deposition of sediments on a
comparatively even surface. The Devonian strata sh slight undulations in the sections exposed on Athabaska river, b .: trend of the
undulations has not been determined and it is doubtfu. ..nether there is
any uniformity of trend. On Peace river and the Slave, where these
rocks overlie beds of gypsum of Silurian age, there is considerable local
disturbance. The beds immediately overlying the gypsum are fractured
and brecciated, and are thrown into a series of anticlines and synclines
sometimes with dips as high as 70 degrees. The disturbance is believed to
be due not to mountain-building forces but to the alteration of beds of
anhydrite to gypsum and the consequent expansion as . result of that
alteration. A gentle anticline crosses Great Slave lake from Pine point to
Nintsi (Windy) point. In the lower part of Mackenzie river the Devonian rocks of the Great Central plain have been considerably influenced
by the disturbances which elevated the Cordillera. The Franklin range

---

1 McConnell, R. G., Geol. Surv., Can., vol. IV, pp. 119D-120D.

has been thrown up as an anticlinal ridge and certain isolated hills, such as Bear rock near Norman, have been produced by faulting. Other fault blocks and anticlinal ridges are described as occurring in the upper part of the Peel River basin. These ridges and fault blocks are related to the Cordilleran disturbances and their trend conforms in a general way to the trend of Mackenzie mountains.

The Cretaceous and younger rocks of the Great Central plain in the southern part of the basin are apparently undisturbed and in the sections exposed on Peace and Athabaska rivers the dip of the strata, except in the foothills, does not exceed a few feet per mile. The hills of Cretaceous rocks which rise out of the country between the Peace and Athabaska rivers, for example, Birch mountain, Buffalo Head hills, Pelican mountain, and Martin mountain, are believed by McConnell[1] to be due to erosion rather than to upwarping.

In the northern part of the basin parts of the Cretaceous of the Great Central plain have been involved to some extent, as has been the Devonian, in the Cordilleran disturbances, and the beds are arched into low anticlines striking north and south and broken by faults. The greater part of the beds here, as in the south, lie, however, in an almost horizontal attitude.

Structure is naturally more evident, if not more pronounced, in the Cordillera than in either the central plain or the Laurentian Plateau portions of the basin. The Rocky mountains, which extend northward along the western edge of the basin as far as the Liard river, are built up of sedimentary rocks that have been elevated by pressure exerted from the west into a series of parallel ridges which strike about north 30 degrees west. These ridges are made up of closely folded and faulted beds sometimes overthrust one over the other, and dipping at high angles generally towards the west. Nearer the central plain in the foothills, the beds are not so closely compressed and the pressure has produced more open folds with lower dips, but with, however, the same general strike.

The Rocky mountains die out in the latitude of Liard river and that stream flows around their northern extremity, cutting through only a few of the more persistent ridges. The beds are here not so compressed and the structure of the ridges is anticlinal.

In Mackenzie mountains "the structure is characterized by folding, generally on a broad scale, which has thrown the strata into a series anticlines and synclines; but the folding is sometimes close, and in certain cases the folds appear to be overturned and overthrust."[2] The general strike of the beds in these mountains as far north as latitude 65 degrees north is slightly west of north, but at that latitude the trend of the ranges and they swing in a wide circle to the westward round the upper of Peel river. On Wind river the mountain building forces have intense and the structure is characterized by open folds with compa low dips to the beds. The strike of the ranges is here east and west

Richardson mountains, which separate Peel river from the Porc have a north and south trend. They have on the whole an anticlinal str ture, and are broken by longitudinal faults. On the eastern side they p sent a steep fault face to the delta of the Mackenzie, and the beds undul gently. In the middle of the range the dips increase to 70 degrees maintain an easterly direction, and on the western side the beds gradually

[1] Geol. Surv., Can., Ann. Rept., vol. V, pt. I, p. 44D.
[2] Keele, J., "A reconnaissance across the Mackenzie mountains," Geol. Surv., Can., 1910.

ECONOMIC GEOLOGY.

BITUMINOUS SANDS.

General Description.

An immense body of bituminous sands, the so-called tar sands, is exposed on Athabaska river. It extends from near Boiler rapid, about 200 miles below Athabaska, to a point many miles below McMurray. Outcrops occur also in the banks of the rivers, and small streams tributary to the Athabaska between these points.

The bituminous sands vary in colour from grey to dark brown or jet black, according to the quantity of contained bitumen and depth of weathering Where they are heavily saturated they are much softened by the heat of summer, and the bitumen issues from the sands and forms pools. It is believed that they consisted originally of unconsolidated sands and soft sandstone ranging in texture from a fine silt to a coarse grit, and induration has been effected by impregnation with bitumen. The formation contains occasional lenticular beds of ironstone and quartzite and thin seams of lignite. The lower part of nearly all exposed sections consists of unstratified sands and the resulting bituminous sands are fairly homogeneous (Plate XIIA). "In passing upward, however, narrow bands of sandstones and occasional quartzites are found interbedded with the originally uncompacted sands (Plate XIIB). These non-bituminous strata gradually increase until by their preponderance they entirely replace the bituminous sands."[1] The formation is overlain by the Clearwater shales, from the first exposure at Boiler rapid down to McMurray. It constitutes the basal member of the horizontally disposed Cretaceous series in this region, and rests directly on the Devonian limestone with very little unconformity to mark the great lapse of time intervening between the deposition of the two. The bitumen with which it is impregnated is believed by many to have had its origin in the Devonian formation.

"Other occurrences in the province of Alberta have been recognized near Bonnie Glen (NW. ¼ sec. 14, tp. 47, range 27, W. 4th mer.); Nakamun (NE. ¼ sec. 28, tp. 56, range 2, W. 5th mer.); Legal (secs. 28 and 32, tp. 57, range 25, W. 4th mer.); Westlock (SE. ¼ sec. 5, tp. 60, range 26, W. 4th mer.); and elsewhere. At none of these localities has bituminous sand been found in commercial quantity, although it is only fair to say that as yet no systematic prospecting has been seriously undertaken. The deposits are, however, so situated that no great outlay would be required to finally determine their commercial value."[2] Bituminous sands are also reported to occur on Wabiskaw river at Prairie river. Fragments of bituminous sand are found at the Upper narrows on Buffalo lake, Saskatchewan. On the east side only small fragments were found; they are low grade and evidently much altered by water action. On the west side bituminous sand occurs as large masses of float, the largest weighing possibly 5 to 8 tons. The beds from which these were derived were not found.[3]

---

[1] Ells, S. C., "Preliminary report on the bituminous sands of northern Alberta", p. 6.
[2] Ells, S. C., "Preliminary report on the bituminous sands of northern Alberta," p. 3, footnote.
[3] Ells, S. C., Mines Branch, Dept. of Mines, Can., Sum. Rept., 1914, p. 64.

S. C. Ells, of the Mines Branch, Department of Mines, Ottawa, has carried on investigations of the bituminous sands of the Athabaska River section with a view to ascertaining their adaptability to commercial ends.[1] The following information has been abstracted from his preliminary report.

"Although the area represented by actual out   s has   been accurately determined, it is probably not less than 750 square mi[2]   Extensions of the deposit under heavy cover, particularly to the   , will greatly increase this estimated area.

"At various points wide variations occur in the quality of the material, the thickness and character of the deposits, and, in those topographical and geographical conditions which must, to a large extent, control possible future development."

Measurements were made of a great number of outcrops, but in many instances, earth slides, the growth   timber along the upper part of an exposure, and the presence of talu   the foot of the slope partly obscured the outcrop, so that only approxi   ons were made of the thickness of the formation and of the covering,   as possible, however, to eliminate many outcrops from further consideration on account of the depth of the overburden, the low bitumen content of the deposit, variability in the bitumen content, variability in the mineral agregate, or difficulties in transportation. The estimated thickness of the bituminous beds and the overburden, and the conclusions arrived at are based wholly on surface indications. Extensive stripping and other systematic exploration would   oubtless modify to a certain extent the figures and conclusions.

Detailed Description.

Between Boiler rapid and Cascade rapid "bituminous sands are probably more or less continuous along both sides of the river, though the actual outcrop is frequently obscured. Such exposures as do occur are usually much banded, and much of the bituminous sand itself is of low grade." There is also a great thickness of overburden.

Between Cascade rapid and McMurray examination was made of a number of   ions exposed at bends in the river. It is probable that beds of workable   e and of commercial quality occur in all these sections, but the overburden is great. Systematic exploration of the areas of less abrupt topography, lying between the exposed sections, may reveal commercial deposits, but such exploration would be expensive. "Therefore, in view of the more favourable conditions under which bituminous sand is found elsewhere in the McMurray district, it is doubtful if, with one or two possible exceptions, the deposits referred to in Section No. 1 (between Boiler rapid and McMurray), should be seriously considered at the present time."

Along the Athabaska, north of McMurray, thirteen of the more promising outcrops were examined in some detail. "Although all are outcrops of what is apparently one continuous deposit, there is, as elsewhere, considerable variation in quality of material, and mode of occurrence. Consequently it is probable that of the thirteen outcrops examined, quite 50 per cent may, for the present, be eliminated from further consideration. The following table indicates measurements of representative sections, and requires no comment."

[1] "Preliminary report on the bituminous sands of northern Alberta". Mines Branch, Dept of Mines, 1914.

Principal Exposures, Section 2. Athabaska River, North of McMurray

Number	Average exposed thickness of high grade	Average exposed thickness of low grade	Total exposed thickness bituminous sand	Overburden	Horizontal distance from foot of outcrop to point at which thickness of overburden is estimated	Lower limit of bituminous sand, above or below water level	Side of river	Approximate length of exposure	Miles north of Forks	Remarks
12 (a)	65		65	115	425	Above.	E.	4,980	2½	Between Poplar island and McMurray: a, b, c, are equidistant sections along outcrop. It is difficult to estimate the thickness of commercial grade sand. The deposit is largely made up of impure partings of clay, etc., which to a great extent destroy its value.
(b)	30		30	165	570	"	"	4,980	2½	
(c)			60	170	440	"	"	4,980	3	
13 (a)	30		90	100+	315	"	"	3,400	4½	Opposite north end Big Poplar island. Bituminous sand is mostly of low grade and much banded. From surface indications, the greater part of it is of no value.
(b)	40		90	190	250	"	"	3,400	6	
(c)			20+	145	330	"	"	3,400	6	
(d)	60		61+	90	345	"	"	3,400	6	
14			80	110	520		"	950	11	Opposite Stony island the exposed part at this outcrop shows many impure partings. The lower part of this section may contain high grade material, but is obscured by talus pile 130 feet up the face. On surface indications this outcrop is of no value.
15			75	70+	625			3,000	12	This exposure is largely covered by talus and drift, and the average content in bituminous sand apparently does not exceed 5 per cent.
16 (a)			90	160	700	10ft above	W.	1,775	16	One and a half miles . . . like lower Twin islands. Surface indications at sections a and b show a low grade deposit of very little value.
(b)			51		575		"	1,775	16	
17			5	15	125	5 ft above	E.	300	24	Surface indications show material of low grade and of little value.
18			30	70+	250	"	"	250+	37	Surface indications indicate this to be of a noncommercial value.
19 (a)	20		65	75	315	"	"	1,225	37½	Much of this exposure is obscured by talus and drift. Surface indications are promising.
(b)	30		60	60+	345	"	"	1,225	37½	
20			30		Sand.			1,500	38	Material shows numerous partings of clays, etc., and surface indications do not indicate deposit of commercial value.
21 (a)	14		14	7	450	Sand.		900+	40	Surface indications promising.
(b)	16		16	3-8	450	Above.		900+	40	Overburden light.
22			35	30+	350	At.		190+	51½	Outcrop highly banded with interbedded impurities and is of little apparent value.
23			20	23	1,010			500	53	North of Pierre au Calumet creek. Highly banded. Surface indications indicate low grade material of little value.
24 (a)			4	65	1,295	Below.		1,200	57½	Some interstratified partings, but surface indications promising.
(b)		45	35	110	1,575	"		1,200	57½	
25			45+	60	950	"		600+	19	Opposite Tar island.
26			60	70	550	"	W.	5,000	52	Just above mouth of Calumet river. Material much banded by impure partings and apparently of no commercial value.
27			70	10-20	200	"	"	500+	36	Two miles below mouth of Red river. Material much banded by impure partings. From surface indications deposit is of no commercial value.

Outcrops of bituminous sands occur in the lower courses of many of the tributaries of the Athabaska. Difficulties of transportation would, however, militate against the development of numbers of these.

On Horse creek, in the lower 5½ miles of the valley, exposures are seen that indicate the presence of a very large tonnage. "Much of this, judged by material seen in outcrops, and apart from grading of mineral aggregate, is of commercial grade; but probably not more than four or five outcrops are so situated as to permit of practical development. The areal extent even of these, is limited by the narrow and precipitous nature of the valley."

There are difficulties in the way of transportation. "At alternate bends in the stream, where the current impinges against the banks, the possibility of heavy clay slides must be taken into consideration. Points opposite such cut banks are, however, usually low and free from the effects of slides."

On Clearwater river, which flows through a larger valley than the Athabaska, only one exposure is seen for 20 miles above the mouth, although there is little doubt that the bituminous sands extend many miles up the river and are obscured by drift and forest growth. The outcrop occurs on the north shore a quarter of a mile west of the mouth of Hangingstone creek. The maximum thickness exposed is 20 feet, and the bitumen content and the grading of the mineral aggregate varies greatly.

On Hangingstone creek, which occupies a deep, notch-like valley, there are outcrops at nearly every bend along the first 5 miles. It may be that by careful exploration some points will be found at which quarrying operations can be carried on successfully.

On Christina river, twenty-four exposures were noted, but measurements were made of only seven, the others being eliminated owing to heavy overburden, prevalence of clay slides, and the presence of impure partings in the bituminous sand itself. The depth of the overburden would be a serious handicap in the development of deposits along this river. "Topographically, the valley of the Christina river resembles that of Horse creek. Its course is, however, more direct and the small areas of bottom lands are consequently fewer.............it is probable that better grades and a better alignment can be secured than on either Hangingstone or Horse creeks. Owing, however, to the unstable and often precipitous character of the banks, it is very doubtful if, even here, a surface line could be maintained within the valley itself."

On Steepbank river, which enters the Athabaska 21½ miles below McMurray, a number of promising outcrops occur. Details are given in the following table, concerning nine of the more promising outcrops of the twenty-six seen.

"Based on surface indications, the deposits of bituminous sand on Steepbank river may be considered quite as promising as those seen on any other stream in the McMurray district. Owing to the sandy and gravelly character of much of the overburden, there is a general absence of the heavy clay slides so frequently met with on some of the other streams. Such a consideration is of importance in considering possible development and transportation within a narrow valley."

Principal Exposures along Steepbank River.

Number.	Average exposed thickness, high grade.	Average exposed thickness, low grade.	Total average exposed thickness, bituminous sand.	Over-burden.	Horizontal distance from water's edge to point at which thickness of overburden is estimated.	Lower limit of bituminous sand.	Side of river.	Approximate length of outcrop.	Distance from mouth of river. Miles.	
54			90	15–20	300	20 feet–40 feet above water level.	N.E.	400	2	Shows much banding due to narrow partings of clay, etc.
55 (a)			160	20	250	15 feet–20 feet above	S.W.	800	2	Outcrop shows much banding and does not appear so favourable as others farther up river.
(b)			180	15	250		S.W.	800	2	
56	70	100	170	25	300	At.	N.E.	275	3	Low grade is highly banded, but high grade is of good quality.
57			30–70	10–90	325	10 feet above.	S.W.	625	3½	Material good and conditions relatively favourable.
58	70	80	150	40	300	Near.	S.W.	200	5	
59	75	80	155	20	300	Near.	N.E.	300+	5½	
60	12		12	8	200	Below.	S.W.	300	5¼	
61	75	60	135	20	250	Below.	N.E.	500	10	
62			120	25	300	Below.	S.W.	425	12	Outcrop shows much banding.

On Muskeg river occasional small deposits of bituminous sands occur, but surface features do not indicate that any of these are of commercial value. Only low grade deposits were observed on Pierre au Calumet creek, and none of commercial importance on Firebag river. Exposures occur on Calumet river, but appear to be of no commercial importance, and those seen on Tar river are low grade and of a highly banded character.

A number of outcrops occur on Moose river, which enters the Athabaska from the west 47 miles below McMurray, and some of those seen between 4 and 8 miles up stream are of sufficient promise to warrant further detailed exploration. "The valley of the Moose is for the most part narrow, with banks rising 150-200 feet. At a number of points, however, changes in the course of the river have resulted in the formation of occasional lateral basins and low-lying areas, which may prove of importance should development of the bituminous sands be undertaken here . . . . As on other streams of a similar character transport- ation within the valley itself. will present difficulties. . Owing, however, to light overburden causing fewer clay slides. the maintenance of a line would not be so difficult as on most of the other tributaries of the Athabaska river."

On McKay river sections of the bituminous sands were seen along the lower 22 miles of its course. Nine were examined in detail; these lie at distances of $\frac{3}{4}$, 2, $2\frac{1}{2}$, $7\frac{1}{2}$, 8, 9, 10, 12, and 13 miles from the mouth of the river. Surface indications lead one to suppose that some of these are of commercial value. "The river flows through a narrow valley, marked by precipitous slopes and high cut banks" and difficulties of transportation would be encountered.

Utilization.

Extraction of Bitumen from Bituminous Sands and Sandstones. "At various localities in the United States during the past twenty years, the commercial extraction of bitumen from bituminous sands and sandstones and from bituminous limestones has been attempted" and many hundreds of thousands of dollars have been spent in the construction of plants for this purpose.

"Generally speaking, commercial extraction in the past has been attempted by the use of solvents—principally carbon disulphide, and the lighter petroleum distillates—and by the use of hot water and steam. Of the first two solvents, carbon disulphide is more expensive and more volatile, while the escaping fumes are a menace to the health of employees. In actual commercial practice, however, it appears that neither the use of naphtha nor of carbon disulphide has been successful. Apart from attendant danger from fire and explosions, there is a serious loss in the solvent employed. Such a loss, due in part to evaporation and in part to failure to fully recover the solvent from the extracted bitumen and from the sand tailings will, at times, probably aggregate nearly 15 per cent.

"The results when hot water and steam have been used have been more encouraging. A fairly rapid and inexpensive separation has been possible, but in actual commercial practice the extraction has not been sufficiently complete. Summarizing all evidence available to the writer. it appears that, as at present understood, the use of hot water or steam. or a combination of the two, will not give a commercial extraction of more than 60 per cent of the bitumen contained in average bituminous sand-rock. In attempting to secure a higher percentage of extraction, a dispro-portionate increase in cost will probably result.

In conclusion Ells says:

"It is thus safe to say that, even apart from the actual merits of any of the processes that have been used, the extraction of bitumen from bitu-minous sand-rock has not met with commercial success. Nevertheless, in view of the various factors that must be taken into account in consider-ing past attempts, it is difficult to say whether, under the most favourable conditions, commercial extraction may or may not be feasible. Meanwhile, those who may care to attempt extraction on a commercial scale and under conditions prevailing in northern Alberta, will have available the results of many years' elaborate and often costly experimentation, on which to base their efforts. Further, it should be remembered that, owing to freight and other charges on imported asphalts, extraction in Alberta would be on a more favourable basis than has, for example, been the case in California, where competing residuum can be sold at a very low figure."

Material for road construction. The applicability of bituminous sands to road construction is dependent on the character of the mineral aggregate, and the proportion and character of the contained bitumen.

Examinations have been made of the grading of the sand in a number of samples and it has been found that in this respect much of the Alberta material is unsatisfactory. "It is, however, possible that a further and more detailed investigation of individual deposits will show that a combina-tion of material from two or more localities will result in a satisfactory grading of the aggregate."

"In general, the per cent of asphalt cement required in various classes of bituminous road construction ranges from 6 to 12. The average per cent of bitumen contained in samples of Alberta bituminous sand examined is about 15. Of this amount 15-17 per cent will be lost in necessary pre-liminary heating. On such a basis the bituminous content actually avail-able in the anhydrous sand will thus be reduced to 12-13 per cent." Some difficulty would be encountered in reducing the proportion of bitumen by the addition of sand. "If, however, the mixer used be furnished with sufficient power and the sand itself be preheated the desired reduction could doubtless be effected............it appears that specially designed machines will have to be constructed before Athabaska bituminous sand can be seriously considered as a paving material...................

"Before it can be successfully adapted to sheet asphalt work, the bitumen contained in the Alberta sand will require a very considerable modification. The penetration of the extracted bitumen is much too high and constitutes the dominant feature in considering its value as an asphalt cement. In the laboratory this feature can be sufficiently modified by proper heating and fluxing. Whether this will be found practicable when undertaken under conditions governing actual paving construction, can best be determined under conditions more nearly approaching those of commercial work."

The "Preliminary report on the bituminous sands of northern Alberta," from which the foregoing information has been abstracted, contains detailed information regarding about eighty different outcrops, analyses, hints on prospecting, and other information of practical value.

During the summer of 1915 some experiments[1] were conducted under
the supervision of S. C. Ells of the Mines Branch, in the practical application of the bituminous sands in street construction in the city of
Edmonton.

"As a site for this pavement, a section of Kinnaird street, immediately
south of Alberta avenue, was selected. The traffic along this part of
Kinnaird street is such as will give a pavement a fairly severe test and may
be classed as heavy. Apart from a considerable volume of fast automobile
travel, it includes vehicles which carry loads up to eight and ten tons."

The shipment for this test consisted of bituminous sands of both
coarse and fine mineral aggregates. Modifications were necessary to
correct such features as high penetration of asphalt cement, unbalanced
mineral aggregates, and excess of asphalt cement (15 per cent).

[2]"The effect of unduly high penetration was modified by partial
distillation of the more volatile fractions. The unbalanced aggregates
of coarse and fine bituminous sand were partially corrected by combining
the two in a proportion of two of fine to one of coarse. In the case of the
sheet asphalt mix, the resulting aggregate was further modified by the
addition of clean and graded sand; in the case of the bitulithic mix, by the
addition of clean sand and graded, crushed gravel. In the case of the
bituminous concrete, fine-grained bituminous sand only was used, and
was modified by the addition of graded crushed gravel and of clean sand.
This manipulation also reduced the somewhat high percentage of asphalt
cement present in the original material to the final percentage desired in
each case."

The pavement was opened to traffic on August 26, 1915. Under date
of November 1, 1916, A. W. Haddow, Acting City Engineer of Edmonton,
reported to Eugene Haanel, Director of the Mines Branch, as follows:

"I have your letter of October 24, inquiring as to the condition of the
experimental pavement which you laid in this city, with the native bituminous sands from Fort McMurray and Fort McKay. I examined this
pavement again yesterday on receipt of your letter, and find it in perfect
condition. The only defect is the transverse crack which is probably due
to contraction of the base and subgrade, and could not be taken in any way
as default of the surfacing.

"The pavement does not show any markings, due to horse traffic in
hot weather. There are no indications of any waving either in the sand
mixture or rock mixture. There is no indication of any pitting, which
we sometimes find in our suburban roads, due to the clay caking on the
surface, and when it flakes off, very often pits the surface by taking a small
portion with it.'

Mr. Ells is of the opinion that with minor modifications in the manipulation adopted in connexion with the experimental paving referred to
above, Alberta bituminous sand   be successfully used as a basis for
satisfactory asphaltic pavements.

"From a comparative study of cost data, based on the use of Alberta
bituminous sand and of imported asphalts, it appears that the application
of the former in the crude state will be restricted within comparatively
narrow limits in western Canada. Indeed, extensive development of the
McMurray deposits will probably depend on the commercial application

[1] Mines Branch, Dept. of Mines, Can., Sum. Rept., 1915, pp. 67-74.
[2] Ells, S. C., Can. Min. Jour., vol. XXXVII, No. 3, pp. 73-74, Feb. 1, 1916.

of an extraction process whereby the bitumen can be marketed in a more or less pure form. Such a process would doubtless ensure for the McMurray product a wide market, not only as a paving material, but in many other recognized applications for high grade bitumen."

Topographic maps embracing many of the most important deposits have been published by the Mines Branch, Department of Mines, Ottawa, on a scale of 1,000 fe   to 1 inch and with contour intervals of 20 feet.

CLAYS.

Investigations have been made of some of the clay deposits on Athabaska river and its tributaries, and of some of the clays and shales along the line of the Grand Trunk Pacific railway.

In the Athabaska district the clays examined immediately overlie the Devonian limestone or are associated with the McMurray sands. " An excessive percentage of carbon is noted in the case of certain of the samples examined. Where the clay lies between the bituminous sand and the Devonian limestone, this contained carbon has been largely, if not altogether, derived from the overlying bituminous sand. It is probable that such contamination would materially decrease on working in from the outcrop." The depth of the overburden, and transportation difficulties are matters for serious consideration. Results of laboratory tests of some of the more important clays are extracted from a report published by the Mines Branch, Department of Mines, Canada, and entitled " Notes on clay deposits near McMurray, Alberta," by S. C. Ells.

" *Laboratory No. 190.* From point on northwest shore of Muskeg river, between head of portage and mouth of river.

" A light grey, very plastic clay, with good working and drying qualities. It burns to a cream coloured, dense, steel-hard body at cone 3, with a total shrinkage of 9 per cent, and softens when heated up to the temperature of cone 27. This is a good example of a stoneware clay, and is also a fire-clay. It is the most refractory clay at present known to occur in the province of Alberta.........

" *Laboratory No. 310.* This clay requires 23 per cent of water to bring it to the best working consistency. It is very plastic, and smooth. The drying must be done slowly after moulding, to avoid cracking. The drying shrinkage is about 7 per cent. The results obtained in burning are as follows:

Cone.	Fire shrinkage. %	Absorption. %	Colour.
010	0	10	buff
06	1·4	8	"
03	3·0	3	"
1	3·4	2	dark buff
5	2·0	0	grey
15	Fused.		

" This is one of the better grades of clay, with good working qualities, and shrinkages within commercial limits. It would be suitable for the manufacture of hard burned fireproofing buff face bricks, or sewer-pipe. The drying qualities could be improved by the addition of a small percentage of ground-burned clay to the raw clay. It must be burned slowly." This sample was taken on McKay river, 11·2 miles from the mouth.

Owing to slides and talus little can be stated regarding the extent of the deposit and transportation to the Athabaska would present difficulty.

Samples 191, 313,314, and 317 were taken along Moose river at distances of 3·3, 6·75, 6·7, and 1·8 miles respectively from the mouth. The deposits are greatly concealed by slides and talus.

"*Laboratory No. 191.* From Moose river, interbedded between bituminous sand and Devonian limestone.

" Dark grey, very plastic, smooth, fine-grained clay of the stoneware type. Burns to a salmon coloured dense body at cone 3, with rather high shrinkage, and fuses at cone 18.

"*Laboratory No. 313.* This clay only required 14 per cent of water for tempering, owing to the presence of a large percentage of rather fine-grained quartz sand. The plasticity and working qualities were low for this reason.

" The drying shrinkage was only 3 per cent.

" The following results were obtained on burning:

Cone.	Fire shrinkage.	Absorption.	Colour.
010	0	8	salmon
06	0	8	buff
03	0	8	"
5	0	7	"
14	begins to soften		grey
18	fused		

" As this clay is too sandy to use alone, a mixture was made by adding 50 per cent of a fat clay (315) from a near-by locality. This gave a body with properties intermediate between the two extremes of a highly plastic clay with large shrinkage, and a lean clay with low shrinkage, so that the results obtained in working and burning were good. The air shrinkage was about 5 per cent. A steel-hard, practically non-absorbent body was produced at cone 1.

" This mixture would probably be suitable for sewer-pipe, or electrical conduits.

"*Laboratory No. 314.* This is a soft, grey clay, with good plasticity and working qualities. Wares moulded from it will stand fast drying without checking.

" The drying shrinkage is 6 per cent.

" The following data were obtained on burning:

Cone.	Fire shrinkage.	Absorption.	Colour.
010	0	12	salmon
06	1·0	9	"
03	2·3	5	"
1	3·4	1	buff
5	4·6	0	grey
9	intact		
14	softens		

" This is a good material, the shrinkages are low, and it gives no trouble on burning. It would be useful for the manufacture of face brick, fire-proofing, electrical conduits or sewer-pipe............

"*Laboratory No. 317.* Light grey clay, with slightly reddish tinge, requiring only 17 per cent of water for tempering. It is rather stiff in working when wet; the plasticity is good, and the clay is very smooth. The drying qualities were not tested, but they are probably good, owing to the small amount of water required to bring it to a working condition.

"The drying shrinkage is 5 per cent, and the following results were obtained on burning:

Cone.	Fire shrinkage.	Absorption.	Colour.
010	0	10	salmon
06	0	10	"
03	1	7	buff
1	1	6	"
5	2	3	grey
9	4	vitrified	"
16	fused		

"This is a stoneware clay suitable for the manufacture of pottery, crocks, jars, teapots, etc. It would require some experimental work to fit the bodies with suitable glazes, but it is probable that the usual Bristol and slip glazes used for stoneware articles would answer.

"Some 3-inch, round, hollow tile, was made on a hand press, and sent to a commercial sewer-pipe plant for a salt-glazing test.

"Salt Glazing Test. The results of the salt glaze tests on this clay show that the glaze cannot be successfully applied at cone 3, as that temperature is too low. The body showed no sign of vitrification, being stil, porous and rather soft.

"It would require a temperature in the kiln of at least cone 5, or better, at cone 6, to produce a glaze on this clay. It will then show a glaze equal to number 315, but of lighter colour. The commercial kiln in which the tests were made, did not give a higher temperature than cone 3, hence it was impossible to complete the test on this clay............

"Laboratory Nos. 319 and 320, are from the point near which the southerly boundary of the Murphy bituminous-sand claim meets the east shore of the Athabaska river. Owing to slide and talus, the thickness of the clay could not readily be accurately determined, but appears to be quite 20 feet. At the points from which the samples were taken the over-burden appears to consist of from 10 to 20 feet of low grade bituminous sand. What appear to be extensions of the same bed reappear along the river at intervals for one-third of a mile to the south of the point where samples were secured.

"Laboratory No. 319. A light grey, highly plastic, and smooth clay, with good working properties. It must be dried slowly, being liable to check if dried too fast. The drying shrinkage is 5·5 per cent.

"The following results were obtained on burning:

Cone.	Fire shrinkage.	Absorption.	Colour.
010	0	11	salmon
06	0	10	"
03	0·6	8	buff
1	1·3	6	"
5	2·0	5	grey
9	4·0	vitrified	"
17	fused		

" This is a typical stoneware clay, suitable for the manufacture of all classes of stoneware articles, and pottery.

" It is not a fire-clay, but may be sufficiently refractory for stove linings, boiler-setting blocks, or other purposes where extremely high temperatures would not be used.

" This is one of the best clays of the series; it closely resembles No. 317, which is almost as good. It is not so refractory as No. 190, which stands up at the highest temperature of any of these clays."

Tests have been made of shales and clays occurring along the line of the Grand Trunk Pacific railway in the vicinity of coal seams west of Edmonton and some of these have been found suitable for common brick, fireproofing, and field drain tile. The results of the tests are published by the Geological Survey, Ottawa, in memoirs written by Heinrich Ries and Joseph Keele.

Ries says that " one of the most promising areas in which to prospect for shales to be used in the manufacture of clay products is in the Jasper Park region. Here, on the south side of the river, several coal seams have been opened up, and associated with them, so far as known, are carbonaceous, sandy shales, which do not appear very promising, but the section cannot be said to have been fully explored, so that other shales in these Cretaceous beds may be developed later."[1]

COAL.

Coal is one of the most important mineral resources of the Mackenzie basin. It occurs in the Cretaceous formations of the southern part of the basin and in the Tertiary formations of the northern part. Mining operations are being conducted on important seams in the Cretaceous rocks along the Grand Trunk Pacific railway, and descriptions of these are found in the publications of the Geological Survey. In this report, however, descriptions will be given of only the more remote and less known seams, which lie farther north.

In the foothills north of the Grand Trunk Pacific railway a coal field extends from tp. 49, range 27, W. 5th mer., northwestward to tp. 59, ranges 7, 8, and 9, W. 6th mer., and beyond. The seams occur in the Kootenay formation. The coal is high-grade bituminous and at least one seam on Smoky river may be classed as anthracite. Exposures occur on Hay river, Smoky river, Sheep creek, and other streams.[2] MacVicar gives analyses of samples taken from a number of seams in different parts of the field. Those from the Isenberg claim, Smoky river, tp. 58, range 8, W. 6th mer., serve to illustrate the range of composition of coals from this field.

[1] Geol. Surv., Can., Mem. 47, 1914, p. 42.
[2] Geol. Surv., Can., Sum. Rept., 1916

Analyses of Coals from Isenberg Claim, Smoky River.

Thickness of seam, feet	Moisture	Volatile matter	Fixed carbon	Ash	B.T.U.
8	0.5	19.6	63.4	16.5	12718
9	0.4	21.1	69.8	8.2	14040
17	1.1	18.4	74.0	6.5	14100
12	0.4	18.5	69.4	11.7	13600
9	0.3	19.8	73.0	6.9	15070
7½	1.3	12.5	78.2	8.0	13882
7	6.9	13.4	81.7	4.0	14706
7	2.9	14.8	80.1	2.2	13800
10	1.7	18.1	73.9	6.3	13990
15	1.6	19.4	71.6	7.3	13255
5	0.9	14.7	82.5	1.9	14987
10	0.7	22.6	72.3	4.4	14800
7½	0.5	18.5	72.8	8.2	14300
10	0.7	15.3	76.2	7.8	13913
13	0.5	18.4	73.5	7.6	14220
9	1.0	20.3	71.9	6.7	14160

Lignite seams 2 or 3 feet thick are found in the McMurray sandstone and a seam 4 feet thick occurs in this formation on Christina river 5 miles above its mouth.[1] Seams of lignite, varying from a few inches to 5 feet in thickness, are exposed at intervals also through the Grand Rapids sandstone on the Athabaska. The quality of the coal, however, is usually inferior.[2] A seam lying just above the Grand Rapids sandstone out-crops at various points through a distance of 15 miles on both sides of the Athabaska, above and below Grand rapid. The seam is 3 to 15 feet thick, but carries numerous clay partings. There are thin bands of fairly clean lignite from 1 to 2½ feet thick.[3] The formation of the Peace River district is coal-bearing. A great many seams are exposed in the canyon a few miles above Hudson Hope, and in the small streams of the vicinity, but only a few exceed 3 feet in thickness.[4]

The formation has been tilted by the uplift forming the mountain around the end of which the river swings at this point, so that to the south and southwest of the mountain the strata dip southwesterly at angles of 10 to 25 degrees. In spite of this uplift the strata are remarkably free from disturbance, with the exception of a fault on Gething creek and a few minor folds.

In Peace River canyon a great number of small seams occur, but only three were seen to be over 2 feet thick. Of these P1[5] measures 2 feet 11 inches, possesses a strong roof and floor, and is of superior quality.

On Gething creek, twelve seams were observed, of which only five are over 2 feet thick and only one, G12, over 3 feet.

[1] McConnell, R. G., Geol. Surv., Can., Ann. Rept., vol. V, p. 39 D.
[2] McConnell, R. G., Geol. Surv., Can., Ann. Rept., vol. V, p. 29 D.
[3] Ells, S. C., Mines Branch, Dept. of Mines, Can., Sum. Rept., 1914, p. 85.
[4] Galloway, C. F. J., Ann. Rept., Minister of Mines, B. C., 1912, pp. 118-136.
[5] Mr. Galloway's numbering of the seams is retained

Sandstone roof
  Shale...
  Coal (dull)...
  Coal (bright)...
  Parting (shale)...
  Coal (bright)...
Sandstone floor.

Section of G 19.

	Feet	Inches
	0	7
	1	4
	1	4
	0	1
	1	0

A seam, which is probably G12, is seen from a point opposite the islands in the cliff on the south side of Peace River canyon for a distance of nearly 2 miles. The analyses from G12 show it to be of superior quality.

Section of M1 and M2 on Moose Bar Creek.

		Feet	Inches	
Sandstone roof.				
Shale...		0	0	
Coal (dull)		0	9	
Shale...		0	7	M 1
Coal (dull)		0	7	
Sandstone...		1	3	
Coal...		0	1	
Sandstone		0	3	
Coal...		0	3	
Shale		2	7	
Coal (bright)		1	11	
Sandstone...		0	4	
Coal (bright).		1	6	M 2
Shale...		0	3	
Coal (bright)		0	1	
Hard shale floor.				

On Johnson creek twenty seams are exposed, but only four exceed 2 feet in thickness and three exceed 3 feet.

Section of J12 and J13.

	Feet	Inches	
Shaly sandstone			
Shale...	1	0	
Coal (dull)......	0	10	
Shale.........	1	0	J12
Sandstone....	0	2	
Coal (dull)........	0	6	
Shale...	2	6	
Coal (bright)...	2	0	
Shale......	0	2	J13
Coal (bright).	1	3	
Sandstone..	0	1	
Shale floor.			

Section of J14.

	Feet	In.ies
Sandstone.		
Grey shale........	1	6
Hard shale....	0	2
White sandstone	0	4
Hard, grey shale......	1	6
Coal.....	1	8
Sandstone......	0	2
Coal.......	2	3
Sandstone floor.		

Section of J15 and J16.

	Feet	Inches	
Sandstone.			
Coal.......	1	3	J15
Sandstone........	1	6	
Shale...	3	0	
Coal (hard),......	2	6	
Shale........	0	1	
Coal.....	0	2	J16
Shale.........	1	0	
Coal.........	0	8	
Shaly sandstone floor.			

Section of J 30.

Sandstone roof		Feet	Inches
Coal		2	3
Coal		0	6
Shale	4 in. to	1	0
Sandstone floor			

It is possible J13 and J14 are identical and correspond to M2. "The sections of J13 and M2 are very similar, each having a small rider of dull coal above it, and the correspondence of these two is highly probable. In the case of J14, however, the similarity is much less, and its analysis would almost exclude the possibility of its being correlated with M2. Unfortunately no samples were taken of J13.

" In cases where a seam is visible over a considerable distance, the thicknesses of the individual benches of coal and shale are very variable, the shale-sandstone partings in the seams being frequently of a lenticular nature, increasing within a distance of a few feet from 1 inch to over 1 foot in thickness, and diminishing again as rapidly. The sections given represent, as nearly as could be observed, average conditions".[1]

Coal seams are also reported as outcropping on Eightmile creek 7 or 8 miles from its mouth.

Analyses of Peace River Coals.[2]

Seam.	Thick- ness.		Hygro, water.	Vol. comb. mat.	Fixed carbon.	Ash.	Coking quality.	Split vol. ratio.
	Ft.	In.						
G12, top bench (dull coal).	1	4	2·9	15·6	79·4	2·1	None	8·15
G12, lower benches.	2	2	2·8	16·9	77·2	3·1	"	7·61
M1 (dull coal)...	0	10	1·0	14·5	70·6	13·0	"	9·43
M2, top bench....	1	1½	3·0	18·0	73·6	5·4	"	6·88
M2, lower bench.	1	7	1·7	16·3	53·7	28·3	"	6·28
J14, top bench..	1	8	2·7	20·9	67·6	8·8	"	5·93
J14, bottom bench.	2	3	1·8	23·9	67·8	6·5	Fair	5·80
J16, top bench...	2	8	1·6	15·9	77·4	5·1	None	8·93
J16, bottom bench	0	8	1·0	24·4	73·7	3·9	Fair	7·21
J20........	2	9	1·4	16·0	73·1	9·8	None	8·91
P1 (Section 1)....	2	11	2·2	15·6	80·6	1·6	"	8·84

Thin seams have been observed in the Dunvegan formation at a few other points. In the gorge of a small brook entering Pine River South near Table mountain, four seams of bright coal occur in 90 feet of alternating beds of sandstone and shale. These seams in descending order are 6, 8, 24, and 6 inches thick.[3] On the lower part of Coal brook, which flows into the east branch of Pine River South, there are several thin seams, the thickest being 6 inches. Coal also occurs on the east branch of Pine River South. A seam of lignite of inferior quality, 12 inches thick, occurs on Brulé river, near its mouth, about 14 miles below Dunvegan.[4]

[1] Galloway, C. F. J., Ann. Rept., Minister of Mines, B. C., 1912, p. 133.
[2] By H. Carmichael.
[3] Selwyn, Alfred R. C., Geol. Surv., Can., Rept. of Prog., 1875-76, p. 53.
[4] Dawson, G. M., Geol. Surv., Can., Rept. of Prog., 1879-80, p. 118B.

F. H. McLearn states that "Lignite occurs in the upper sandstone member of the Peace River sandstone, approximately 25 feet below the top at a point about 10 miles below Peace River. It is also found at a similar horizon near Peace River on Heart river, etc. Owing to the lateral changes in composition and thickness this coal horizon can hardly be considered from the standpoint of large scale operations. However, where locally the coal is of good quality and of sufficient thickness and is of marketable value not far (a few feet) from the face, i.e., all coal removed can be sold, small scale operations may be advisable. A sample of a layer of lignite 7 inches to 1 foot 2 inches thick, underlain by 1½ feet of lignite shale (ash 58·29 per cent) gave the following analysis (Mines Branch, E. Stansfield, Chief Engineering chemist).

	Per cent
Moisture	11·2
Ash	13·9
Volatile matter	28·9
Fixed carbon	43·0

" The greatest thickness of clean lignite measured in the cliffs by the writer was 1 foot 6 inches. Marketable coal of greater thickness, however, may be present in places.

" The Dunvegan sandstone cannot be looked upon as a coal producer. A small lignite seam is present in the cliffs below Montagneuse river, but is never thicker than 6 inches."

Lignite coal is seen at a number of places in the banks of the Athabaska for 62 miles above Fort Assiniboine and several miles below. About 28 miles above the mouth of McLeod river the following section is found.

	Feet.	Inches.
Sandstone and shales		
Shaly lignite	3	0
Soft sandstone and shale	4	0
Cheel lignite	1	8

Two beds occur about 8 miles farther down[1]. "They are shown near the water's edge in a slide detached from the main bank. The upper seam here has a thickness of ten feet, without including in this measurement about six thin, shaly partings, which make up in all about ten inches of shale. Below this seam is about twenty feet of soft, earthy sandstone and shale followed by a second seam of clean, hard, lignite coal three feet in thickness." G. C. Hoffmann[2] gives the following analyses.

	Ten-foot seam.	Three-foot seam.
Hygroscopic water	11·47	10·58
Volatile combustible matter	32·09	32·79
Fixed carbon	47·79	50·19
Ash	8·65	6·44

Two thin seams of lignite occur at what is probably the same geological horizon farther down the river.

Fragments of coal were observed on the upper course of Wapiti river and on its tributary, Mountain creek, and thin seams occur in the banks of the latter. Drift lignite is also abundant in Little Smoky river.

Coal occurs on McLeod river and near Wolf creek in tp. 52, range 15, W. 5th mer.[3] Thin seams have been observed on Fort Nelson river

[1] Dawson, G. M., Geol. Surv., Can., Rept. of Prog., 1879-80, p. 126B.
[2] Geol. Surv., Can., Rept. of Prog., 1879-80, p. 10H.
[3] Dowling, D. B., Geol. Surv., Can., Sum. Rept., 1909, p. 149.

and Pine river,[1] and some outcrops occur on Doig river where it crosses the twenty-third base line.[2] Lignite has also been observed in township 78 along the fifth meridian[3] and there are thin seams at several horizons in the plateau to the south of Lesser Slave lake.[4] Fragments of lignite are found on the lower slopes of Marten mountain and at many other points in the Peace River district. Small seams of impure lignite occur on Liard river, and Coal river, one of its tributaries, brings down fragments of similar material.[5]

In describing the coal deposits of the Peel River basin Camsell states that "Seams of lignite occupy extensive areas in the rocks of the Tertiary basin at the Bonnet Plume river. The largest seam noted was thirty feet in thickness, another was eight feet, and several varied from two inches to ten. The lignite is not of very good quality, and has been burnt in many places by the fires which have been in existence for many years. Lignite also occurs a few miles above the mouth of Cariboo river, and also in the canyon of the Rat river above the mouth of Barrier river. Many sections of the Peel plateau below Snake river show beds of peat resting on the clay or sandstone, sometimes as much as twelve feet in thickness."[6]

The Tertiary beds on the Mackenzie at the mouth of Great Bear river hold several seams of lignite ranging in thickness from 2 to 4 feet,[7] and a seam exceeding 9 feet in thickness was reported by Sir John Richardson as being visible a short distance above Great Bear river, at low water, during the autumn.[8] Drift lignite is also found on the lower part of Gravel river.[9]

Franklin mentions that Garry island at the mouth of the Mackenzie "is terminated to the northwest by a steep cliff, through which protrude, in a highly-inclined position, several layers of wood-coal, similar to that found in the Mackenzie."[10]

R. MacFarlane, in speaking of the Lockhart river below its junction with the Iroquois, says: "The formation of the banks of the Lockhart for some distance after we fell upon it, consisted of a bituminous coal, resting on a bed of limestone with an upper layer of vegetable mould covering a bed of from two to ten feet of clay, underneath which the carboniferous stratum appeared. Lower down, the formation was perceived to be stratified shale."[11]

COBALT.

No important deposits of cobalt are known to occur in the basin, but the rocks on the east shore of McTavish bay, Great Bear lake, are stained in places with cobalt-bloom,[12] and at one locality on the north shore

[1] Ogilvie, Wm., Top. Surv. Branch, Dept. of the Interior, Ann. Rept., 1892, p. 27.
[2] Akins, J. R., Top. Surv. Branch, Dept. of the Interior, Ann. Rept., 1911-12, p. 58.
[3] Ponton, A. W., Top. Surv. Branch, Dept. of the Interior, 1908-09, p. 170.
[4] McConnell, R. G., Geol. Surv., Can., Ann. Rept., vol. V, p. 40D.
[5] McConnell, R. G., Geol. Surv., Can., Ann. Rept., vol. IV, p. 41D.
[6] Camsell, Charles, Geol. Surv., Can., Ann. Rept., vol. XVI, p. 47CC.
[7] McConnell, R. G., Geol. Surv., Can., Ann. Rept., vol. IV, p. 31D.
[8] "Arctic searching expedition—a journal of a boat-voyage through Rupert's Land and the Arctic sea in search of the discovery ships under command of Sir John Franklin," 1851, p. 189.
[9] Keele, Joseph, "Reconnaissance across the Mackenzie mountains on the Pelly, Ross, and Gravel rivers, Yukon and Northwest Territories," Geol. Surv., Can., 1910, p. 50.
[10] Franklin, John. "Narrative of a second expedition to the shores of the Polar sea in the years 1825, 1826, and 1827," p. 37.
[11] Can. Rec. of Sc., vol. 4, 1890, p. 32.
[12] Bell, J. M., Geol. Surv., Can., Ann. Rept., vol. XII, p. 27C.

of the bay west of the narrows between Christie and McLeod bays, Great
Slave lake, cobalt-bloom was observed[1], associated with green copper
stains arising from the weathering of thin plates of chalcopyrite in joints
in greenstone.

COPPER.

No important copper deposits are known to occur in the Mackenzie
River basin. Native copper has been reported, however, at the northeast
end of McTavish bay.[2]

GOLD.

Gold has not been discovered in any important quantity in the
Mackenzie basin east of the Rocky mountains, and the extent of country
known to be favourable for prospecting for this metal is quite limited.

Ogilvie, who descended Peace river from Fort St. John to Peace River
Crossing in 1891, states that "Gold is found in small quantities on Peace
river, and at present there are several miners on that stream."[3] McConnell
in describing gold placers on this river says that "Three miles above the
mouth of Battle river, a large bar nearly a mile long, on the left bank, was
examined, from which we obtained fifteen to twenty colours of fine gold,
by washing a few handsful of the mixed gravel and sand in an ordinary
frying pan. We tried the bar at several points and always with the same
result. A small stream descends from the plateau on the opposite side of
the river, and by leading its waters across the river, which is here about
1,000 feet wide, the bar might be easily and inexpensively worked on a
large scale. Twelve miles farther up the river, another bar was examined,
which yielded from twenty to forty colours, when washed in the same way.
Numerous other bars occur in this portion of the river, which would probably give as good results as those examined.

"The presence of fine gold in some quantity in the bars above the
mouth of Battle river is probably due to the diminution in the strength of
the Peace River current which takes place here, and its consequent loss of
transporting power. The same fact is shown in the gradual substitution
of sand-bars for gravel-bars which occur at the same point.

"Besides the gold on Peace river, two colours were also washed out
of a bar on Loon river, an eastern tributary of the Peace."[4] Small quantities of gold have been obtained also on McLeod river.[5]

The gold occurring in these rivers probably has an origin similar to
that found in the North Saskatchewan. It may be derived in part from
the glacial drift, portions of which have been transported from the Pre-
Cambrian formations to the northeast.[6] Tyrrell, however, in discussing the
origin of the gold of the North Saskatchewan, presents evidence to show
that "the fine gold in the river was derived directly from the Cretaceous
(Edmonton) rocks which form the banks of the stream, and that these
rocks in their turn were an old and very low grade placer deposit which

[1] Bell, Robt., Geol. Surv., Can., Ann. Rept., vol. XII, p. 108A.
[2] Bell, J. M., Geol. Surv., Can., Ann. Rept., vol. XII, p. 27C.
[3] Dept. of the Interior, Ann. Rept., 1892, pt. 7, p. 28.
[4] Geol. Surv., Can., Ann. Rept., vol. V, pp. 62D–63D.
[5] McEvoy, James, Geol. Surv., Can., Ann. Rept., vol. XI, p. 41D.
[6] Dawson, G. M., Geol. Surv., Can., Ann. Rept., vol. XI, p. 15A.

has come originally from the mountains west of the Upper Columbia valley."[1]

Small quantities of gold occur in the bars of Liard river. Mining operations have been conducted between Devils portage and the mouth of Dease river, but no deposits of economic value have been found below Devils portage.[2]

Mining has been carried on at Porcupine bar, and at Bed-rock bar 4½ miles lower down. Both of these are on that section of the river having a general north and south trend, about 25 miles above the mouth of Turnagain river. An auriferous bar occurs at the mouth of Rabbit river, and below Portage brûlé the river is wide and filled with low islands and bars, some of which are auriferous. McCullough's bar, on which gold in paying quantities was first discovered on the Liard, occurs in this vicinity. Colours of gold were also obtained on the Liard at the mouth of Fort Nelson river.

Camsell, in reporting on the Peel river and its tributaries, states that "Some coarse colours of gold were panned out from a shovelful of dirt scraped from the rim rock at the mouth of Little Wind river", but only very fine colours were obtained from the gravels of the mountain section of Wind river. Gold is reported to have been found in the gravels of Hungry creek, but "very little indication, however, of placer gold was found on the bars within five miles of ₊ mouth. As the stream rises in a large lake twelve or fifteen miles up, and flows through a low muskeg country to join the Wind river, it appears to be rather an unpromising place for the occurrence gold, but some of its tributaries which flow through a more hilly country might carry the precious metal."

"In panning for gold on a bar on the Peel river above the mouth of the Wind half a dozen fine colours were obtained, showing that this stream contains more of that metal than the Wind river. Gold is reported to have been found by the Indians in the gravels of the Bonnet Plume river, and some specimens were exhibited; time, however, did not permit us to substantiate this report."[3]

Only a small part of the area underlain by Pre-Cambrian formations has been explored. This part is underlain chiefly by granites and gneisses, and these, wherever they have been examined in this and other parts of the Canadian Shield, have as a rule offered very little encouragement to the prospector.

The areas of altered sediments and basic igneous rocks that are found north of lake Athabaska and north of Great Slave lake offer more favourable ground for prospecting. A sample, weighing 5 ounces and consisting of weathered crystalline dolomite, carrying some iron pyrites and a very little brown zinc blende, was examined by the Geological Survey for Mr. E. A. Blakeney and found to carry 2·158 ounces of gold and 0·408 ounce of silver to the ton. The sample was said to have been taken at a depth of 14 feet from one of a series of claims lying within a radius of 10 miles of the mouth of Yellowknife river, on the north side of Great Slave lake.[4]

The glaciation to which the Pre-Cambrian area was subjected renders the occurrence of important placers highly improbable.

[1] Can. Min. Inst., Bull. 34, p. 80.
[2] McConnell, R. G., Geol. Surv., Can., Ann. Rept., vol. IV, p. 29D.
[3] Camsell, Charles, Geol. Surv., Can., Ann. Rept., vol. XVI, p. 46CC.
[4] Geol. Surv., Can., vol. XI, p. 33R.

111

GYPSUM.

Gypsum is exposed at many points in the Mackenzie basin. The deposits that are most likely to prove of commercial interest are those occurring on Peace river and in the escarpment to the west of Slave river. ¹Gypsum is exposed in cliffs (Plate XIII) on both banks of Peace river, almost continuously for a distance of 15 miles below Little rapids. The bottom of the bed is concealed, but the exposed part varies in thickness from a few feet to 50 feet, an exposure of 50 feet occurring on the south side of the river at the foot of the rapids. The surface features indicate that the bed extends back from the river a considerable distance. The gypsum is usually white and massive, but in places is earthy and thin-bedded and holds narrow bands of dolomitic limestone. Thin veins and beds of satin spar are common and anhydrite is occasionally present in rounded nodules and thin beds. The gypsum is overlain by a fractured and broken bed of limestone, but in many places it reaches to the top of the cliff and is covered by only 5 to 15 feet of drift. The deposit is very favourably situated for quarrying.

²Gypsum outcrops at several points in the escarpment west and south of the brine springs at the forks of Salt river. Four miles south of the springs, thin-bedded, white or greyish gypsum, containing occasional narrow layers of anhydrite or beds of dolomite, is exposed in the cliffs for a length of half a mile and with a thickness of 40 to 50 feet. North of this it appears to decrease in thickness and is overlain by beds of grey, crystalline dolomite. At Little Buffalo river 10 feet of impure gypsum is exposed at the base of the escarpment. At a point in the escarpment 8 miles southwest of Fitzgerald there is an exposure of 20 feet of thin-bedded, white gypsum overlain by 10 feet of dolomitic limestone. The prevalence of sink-holes on the top of this escarpment, lying to the west of Slave river, indicates that this mineral occurs throughout the greater part of its length.

Ten feet of somewhat earthy, thin-bedded gypsum, of white, grey, or bluish colour, overlain by 20 feet of limestone, outcrops on the west side of Slave river a few miles below La Butte. Immediately below point Ennuyeux, on the same river, thin-bedded, impure gypsum 4 feet thick, is exposed near the water-level at a medium stage of the water, and at Bell rock, 7 miles below Fort Smith, gypsum is said to have been struck in an excavation.

The numerous exposures in the Slave River country and the topographical features indicate that the mineral is very widely distributed.

At Gypsum point on the north shore of Great Slave lake and along the southwest shore of the north arm of the same lake, thin seams of flesh-coloured gypsum occur between the bedding planes of calcareous sandstone and arenaceous limestone.³

A bed of gypsum 130 feet thick is said to have been struck at a depth of 590 feet in well No. 1 of Athabaska Oils, Limited, located on Athabaska river 9 miles below McKay.

¹ Macoun, John, Geol. Surv., Can., Rept. of Prog., 1875-76, p 89.
Camsell, Charles, Geol. Surv., Can., Sum. Rept., 1916.
² Camsell, Charles, Geol. Surv., Can., Ann. Rept., vol. XV, pp 159A, 167A.
Geol. Surv., Can., Sum. Rept., 1916.
³ Cameron, A. E., Geol. Surv., Can., Sum. Rept., 1916.

In a section exposed by a small stream in the heart of Bear rock at the mouth of Great Bear river "the lowest beds seen consist of reddish and greenish shales, alternating with layers of pink-coloured gypsum and cut by numerous veins and seams of a white, fibrous variety of the same mineral. The gypsum in parts of the section replaces the shale almost altogether, and the layers are separated by mere films of greenish and reddish argillaceous material. The base of the gypsiferous shales was not seen, but they are at least several hundred feet in thickness."[1] Thin layers of gypsum interstratified with dark grey, shaly dolomite occur in several places on mount Charles, on Great Bear river.[2]

IRON.

No important deposits of iron ore have yet been discovered in the Mackenzie basin.

Hematite is found interbedded with quartzite of the Tazin series on Beaverlodge bay on the north shore of lake Athabaska. The quartzite here forms a syncline with limbs dipping at angles as high as 70 degrees; the syncline pitches to the southwest at an angle of 30 degrees. The total area of iron-bearing formation exposed is about 250 acres. The hematite, which also occurs in cavities and along fracture planes, varies from hard blue to soft red, but was not found in economic quantities.[3]

Fragments of iron ore up to 15 pounds in weight are found at a number of points on Steepbank and Moose rivers, tributaries of the Athabaska, and at a point on Steepbank river 4·0 miles from the mouth there is a compacted bed 1 to 2 feet thick made up of fragments of siderite weighing up to 20 pounds. The bed is overlain by a thin capping of bituminous sand and is underlain by a bed of clay 1 to 4 feet thick, which rests upon Devonian limestone. A representative sample of the iron ore was found to contain 35 per cent of iron and 18 per cent of insoluble matter. So far as observed the deposit has no economic value.[4]

Keele[5] reports hematite as occurring on Gravel river about 10 miles below the mouth of Natla river. "This iron is coarsely laminated with red siliceous slate, having a thickness of from 50 to 100 feet, and is interbedded between conglomerate and dolomite. An assay of an average sample of this ore was made at the assay office of the Mines Branch, and gave only 25 per cent of iron."

A vein carrying fibrous, botryoidal, and micaceous hematite is found at Rocher Rouge, McTavish bay, Great Bear lake; and on some of the more southerly of the group of islands known as Les Iles du Large, Great Slave lake, a granular, schistose aggregate of quartz and micaceous hematite occurs in lenticular veins and stringers in a greenish, siliceous sandstone.[6]

Fragments of magnetite and hematite associated with red jasper are found in the Bonnet Plume and Snake rivers. Similar float is widely distributed over a great part of the Peel River basin.[7]

[1] McConnell, R. G., Geol. Surv., Can., Ann. Rept., vol. IV, p. 101D.
[2] Bell, J. M., Geol. Surv., Can., Ann. Rept., vol. XII, p. 25C.
[3] Alcock, F. J., Geol. Surv., Can., Sum. Rept., 1916.
[4] Ells, S. C., Mines Branch, Dept. of Mines, Can., Sum. Rept., 1914, p. 63.
[5] "A reconnaissance across the Mackenzie mountains on the Pelly, Ross, and Gravel rivers, Yukon and Northwest Territories," p. 50.
[6] Geol. Surv., Can., Ann. Rept., vol. XII, p. 36R.
[7] Camsell, C., Geol. Surv., Can., Ann. Rept., vol. XVI, p. 46CC.

LEAD AND ZINC.

Small deposits of galena and zinc blende occur at a number of points on the south side of Great Slave lake about 9 miles inland from Pine point. The country rock is a porous, crystalline, dolomitic limestone interbedded with a pebbly limestone lying either in a horizontal attitude or in gently undulating folds. Outcrops, however, are infrequent and the whole region is heavily wooded. Two systems of fracture planes traverse the rock striking north 25 degrees east and south 80 degrees east. The deposits are replacements along bedding or fracture planes and consist of coarse galena, light-coloured zinc blende, and some iron pyrites. They are disconnected and small, being only a few inches in thickness and a few feet in length. Claims have been staked on them a number of times and a few shallow pits sunk, but the value of the known occurrences is inconsiderable because of the low content of silver.

Argentiferous galena containing 38-86 ounces of silver to the ton of pure galena is stated to occur at another point in the vicinity of Great Slave lake east of Resolution, but the exact locality is not given.[1] The rock is described as 'an association of grey mica schist with a white subtranslucent quartz, more or less thickly coated with hydrated peroxide of iron, carrying some coarsely crystalline galena.[2]'

NICKEL.

No nickel deposits that are ' ------ -ni value under present conditions have been discovered. Nickelifero   . les, however, occur at the east end of lake Athabaska; and there has b. ' considerable prospecting and a number of claims have been staked. The geology of this region was examined by J. B. Tyrrell[3] in 1892 and by F. J. Alcock[4] in 1914. An examination of the mineral claims was also made by Charles Camsell in the summer of 1915.

"The rocks[5] of that portion of the north shore in which the mineral claims were located, namely from the Paris group near the mouth of Grease river to Camille bay, belong to one or the other of the following groups:

"(1) A complex of highly foliated and often contorted crystalline rocks of a gneissoid character, the oldest formation in that district.

"(2) The Athabaska sandstone, flat lying and undisturbed on the south shore of the lake, but somewhat disturbed and metamorphosed to quartzite on the north shore where it comes in contact with the norite.

" (3) A foliated norite younger than, and intrusive into, the Athabaska sandstone as well as into the gneiss.

' (4) Dykes and sills of diabase which are probably genetically connected with the norite.

" The gneiss occupies the lake shore from Robillard river westward. It is the oldest of the four above-mentioned formations and is of no importance from an economic point of view. It is the typical Laurentian granite gneiss with some dark lenses of basic rock and dykes of lighter

[1] Bell, R., Geol. Surv., Can., Ann. Rept., vol. XII, p. 103A.
[2] Geol. Surv., Can., Ann. Rept., vol. XI, p. 33R.
[3] Geol. Surv., Can., Ann. Rept., vol. VIII, pt. D.
   Geol. Surv., Can., Sum. Rept., 1914, pp. 60-61.
   Geol. Surv., Can., Sum. Rept., 1915, pp. 123-126.

coloured pegmatite. It is much disturbed and contorted, but has a general trend to the northeast.

"The Athabaska sandstone forms the south shore of the lake, but east of Pine Channel narrows it occupies also part of the north shore where it is somewhat disturbed. Its age is assumed to be Keweenawan. On the south shore it is a white siliceous sandstone lying in a horizontal attitude, but on the north shore of the lake, owing to the intrusion of the norite, it becomes a garnetiferous quartzite and has been tilted at a higher angle with dip to the southward. On the contact of the norite it becomes mineralized with iron sulphides and many mineral claims have been staked on it.

"The norite occupies the area lying between the gneiss and the sandstone on the north side of the lake. It extends from the mouth of Robillard river eastward to the end of the lake, but east of Pine Channel narrows it lies a short distance back from the lake shore and is separated therefrom by a strip of metamorphosed Athabaska sandstone. It is distinctly foliated and cut by many small and irregular veins of quartz. The strike of the foliation varies from northeast to east and the dip is southward at an angle of about 40 degrees. It has apparently been intruded in the form of a great sill or series of sills between the gneiss and the Athabaska sandstone, the former below and the latter above. The actual nature of the intrusion of the norite, whether a single sill or a series of sills, has an important bearing on the occurrence of any ore deposits in it, but this was not definitely determined, though the weight of evidence appears to favour its having reached its present form by a succession of intrusions. It is cut by sheets of diabase and is mineralized at intervals along the same horizons by sulphides of iron. These bunches of mineralized norite constitute the principal "showings" of the district and on them the greater number of mineral claims have been staked.

"Diabase dykes and sills intrude both the gneiss and the norite, but they are generally of small size. They contain some calcite in vugs and small fracture planes, but no regular veins were noticed.

Mineral Deposits.

"Practically all the mineral claims that have been located are situated in the norite, and with the exception of the contact phase of the sandstone and some of the dark bands in the gneiss this is the only formation that shows any evidence of important mineralization.

"The norite is a siliceous foliated rock, striking northeasterly and dipping to the southward. At wide intervals on the same strike and at the same horizon in the formation are bodies or bunches of ore weathering red and consisting of pyrrhotite, pyrite, chalcopyrite, and arsenopyrite disseminated through a gangue of silicified country rock. The sulphides are not massive except in small cross factures, 1 to 2 inches wide, which traverse these bodies. The width of these bodies ranges from 5 to 20 feet and the boundaries in this direction are fairly definite. The length is also variable and while most of them are only a few feet in length the maximum of those seen is perhaps 100 feet. Along the strike the sulphides gradually diminish in quantity until they disappear altogether. The bodies are isolated from each other and are not continuous; but by following out the strike of the foliation of the norite another body may be found several hundred feet away. In places quartz stringers 2 to 6 inches wide crosscut the formation and in these the principal sulphide is arsenopyrite with sometimes specular hematite, pyrite, or galena.

" Deposits of this character are found in several places in the belt of norite lying between Sucker bay and Norite bay and extending along the strike of the norite from the islands about Channel point northeastward for several miles inland. Typical deposits of this character are those on the Norah; Victory, Excelsior, Garrett, and North Star mineral claims. The metals found by assay to occur in them are nickel and copper, but the average value is not high. Some platinum is also said to have been found.

" Samples 1, 2, 3 were collected by the writer from the more important localities where some development work had been done and assayed both by the Mines Branch in Ottawa and by J. A. Kelso, director of industrial laboratories in the University of Alberta. Samples 1, 5, 6 were collected by F. J. Alcock during 1914. The results of analyses are as follows:

Analyses of Fond du Lac Ores.

Copper	0.09	Nil		Trace	Trace	0.65
Nickel.	Nil	Nil	0.10	Trace	0.20	1.07
Gold....		Nil	Nil	Nil	Nil	Nil
Silver.		Nil	Nil	Nil	Nil	Nil

1. From the Norah mineral claim, carrying disseminated pyrrhotite and chalcopyrite. Analysed by H. A. Leverin, Mines Branch, Dept. of Mines.
2. From the Norah mineral claim. (Analysed by J. A. Kelso, University of Alberta.)
3. From the Garrett mineral claim, formerly known as the Lake Point. (Analysed by J. A. Kelso, University of Alberta).
4. From the Norah mineral claim. (Analysed in the Laboratory of the Mines Branch, Dept. of Mines).
5. From the Lake Point mineral claim, now known as the Garrett. (Analysed in the Laboratory of the Mines Branch, Dept. of Mines.)
6. Picked sample containing small quantities of pyrite, chalcopyrite, and pyrrhotite from one of the small fractures on the Paris claim.

" On the Paris group of claims, or what was originally known as the Athabaska group, the rock is a dark, fine-grained variety associated with a porphyritic granite gneiss. The rock outcrops in a low cliff on the shore of the lake and is sparingly mineralized with pyrite and pyrrhotite which is disseminated through it and gives a rusty stain to the outcrop. In narrow fracture planes there is a little more iron sulphide, apparently secondarily deposited, but nowhere does there seem to be any concentration of the sulphides sufficient to form an ore-body of workable dimensions. Assays show the principal valuable metal at this point to be nickel.

" At this point native silver was said to have been found in the small fractures traversing the rock, but I could find no evidence of that metal.

" The claims situated at the head of Camille or Fishing bay are of somewhat different character. Here the prevailing rock is a garnetiferous gneiss presumably altered Athabaska sandstone, intruded by sheets of norite. The outcrop of the quartzite is red with iron oxide and constitutes what is known as the lead on which the claims have been staked. The deposits are of a contact metamorphic nature with mineralization by pyrite and pyrrhotite as a result of the intrusion of the norite. The sulphides are disseminated through the gneiss over a considerable area along a well-defined strike, but do not appear to be sufficiently concentrated anywhere to form important ore bodies.

"In conclusion it is the writer's opinion that, while the mineral deposits that have been taken up in the norite may be of value at some time in the future, unless more important discoveries are made the remoteness of the district from railway transportation and from any large centres of population, and the present undeveloped state of the surrounding country render their exploitation impracticable at the present time."

PETROLEUM AND NATURAL GAS.

Attention has been directed for some time to the possibilities of discovering petroleum and natural gas in the Mackenzie basin where numerous striking indications of their occurrence have long been known. Unfortunately our knowledge of the geological structure is too limited to permit of our offering suggestions in detail as to where explorations might be carried on with the greatest likelihood of success. Boring operations have been conducted along Athabaska and Peace rivers where strong flows of gas were struck in two or three wells, and petroleum-bearing strata on Peace river have been penetrated.

Oil and gas indications are found in the sediments of Cretaceous and Devonian age, both of which are of wide distribution.

The most striking surface indication in the section of country underlain by Cretaceous rocks is the enormous deposit of bituminous sand occurring at the base of the Cretaceous sediments on Athabaska river and its tributaries. This bituminous sand is described elsewhere in this report.

A spring of natural gas is situated at Tar island on Peace river about 25 miles below Peace River Crossing. The gas rises with salt water and some tar from among the gravel and boulders at the upper end of the island. The flow of gas has been roughly calculated to be about 3 or 4 cubic feet per minute.[1] On Peace river and around Lesser Slave lake bitumen has been found at a number of places lining cracks in nodules, and a tar spring is reported as occurring near the mouth of Martin river on Lesser Slave lake[2].

Franklin mentions having observed a bituminous liquid trickling down the steep cliff on the northwest part of Garry island at the mouth of the Mackenzie.[3] On the right bank of Peel river, 10 miles below the lower canyon, a 3-foot vertical fissure cutting across sandstone and shales is filled with a light, soft, carbonaceous substance which burns readily with a red flame, leaving very little ash. This substance has its origin probably in the bitumen of the adjacent rocks.[4]

Gas springs are numerous. The most important occurs at the mouth of Little Buffalo river, a tributary of the Athabaska. The gas "issues

[1] Camsell, Charles, Geol. Surv., Can., Sum. Rept., 1916.
[2] Ogilvie, Wm., Dept. of the Interior, Ann. Rept., 1889, pt. 8, p. 94.
[3] "Narrative of a second expedition to the shores of the Polar sea in the years 1825, 1826, and 1827", p. 37.
[4] Camsell, C., Geol. Surv., Can., Ann. Rept., vol. XVI. p. 47CC.

from the surface in numerous small jets distributed over an area fifty feet or more in diameter. Some of the jets burn steadily when lighted, until extinguished by heavy rains or strong wind, and afford sufficient heat to cook a camp meal. A second spring was noticed on the left bank of the Athabaska about thirteen miles below the mouth of Pelican river. The volume of gas escaping here is less than at the mouth of Little Buffalo river...... Escaping jets of gas were also noted at several points farther up the river, but these were mostly small and may possibly be due to decaying vegetable matter."[1]

Sulphur springs are of common occurrence. Among the number observed are those occurring on the left bank of Christina river 12 miles from its mouth,[2] on the north side of Clearwater river 4 miles below Cascade rapid and other points,[3] on Firebag river[4], on Jackfish and Little Buffalo rivers west of Slave river[5], at Sulphur point on the southwest shore of Great Slave lake, north of Brûlé point on the north shore of the same lake[6], and at La Saline on Athabaska river.

It was thought that the bitumen of the McMurray sandstone on the Athabaska had probably been derived from petroleum by the evaporation of the lighter hydrocarbons and by chemical changes and that at some distance from the outcrop this sandstone might carry petroleum. With a view to testing this hypothesis the Dominion Government undertook boring operations at two points on the river where it was known that the sandstone had a considerable depth of covering. Wells were sunk at Athabaska and near Pelican river.

At Athabaska boring was carried on during portions of the summers of 1894, 1895, and 1896, and a depth of 1,770 feet was attained. Great difficulty was encountered on account of the caving of shale. It was believed that drilling ceased in the Clearwater shales. Gas in small quantities was struck at several horizons.

Log of Well at Athabaska.[7]

	Feet.
Drift.	0-14
Grey shale, soft and caving badly...	14-245
(At 23 feet, 136 feet, and 245 feet, hard streaks were met. Below the hard streak at 245 feet a strong flow of gas.)	
Soft shale.	245-400
(A heavy flow of gas at 334 feet, a hard streak at 338 feet).	
Shale, slightly harder. .	400-425
(At 425 feet, a hard stratum about 1 foot thick).	
Grey shale.	425-500
Darker shale, soft, caving badly.	500-550
Shale with streaks of sand rock 1 to 2 feet thick	550-580
Dark shale, very soft. .	580-825
(At 780 feet salt water was struck, and a strong flow of gas).	
Shale, harder and bluer	825-900
Soft, dark shale	900-1,015
Hard, light shale	1,015-1,037
Dark shale.	1,037-1,090
Sandstone, carrying water.	1,090-1,130
Dark shale, caving badly. . . .	1,130-1,170
Dark shale with layers of sandstone. . .	1,170-1,207
Dull reddish shale and sandstone.	1,207-1,233

[1] McConnell, R. G., Geol. Surv., Can., Ann. Rept., vol. V, p. 64D.
[2] McConnell, R. G., Geol. Surv., Can., Ann. Rept., vol. V, p. 39D.
[3] Bell, Robert, Geol. Surv., Can., Rept. of Prog., 1882-83-84, p. 26CC.
[4] Dowling, D. B., Geol. Surv., Can., Ann. Rept., vol. VIII, p. 67D.
[5] Camsell, C., Geol. Surv., Can., Ann. Rept., vol. XV, pp. 161A, 166A.
[6] McConnell, R. G., Geol. Surv., Can., Ann. Rept., vol. IV, pp. 69D, 75D.
[7] Geol. Surv., Can., Sum. Rept., 1894-5-6.

Log of Well at Athabaska.—Concluded.

	Feet.
Dark, soft shale...	1,233-1,237
Light grey shale, very hard.	1,237 '-1,242
Light grey shale, soft	1,242-1,247
Dark shale, soft....	1,247-1,255
Sandstone, very hard.	1,255-1,260
Soft, dark shale	1,260-1,285
Hard sandstone.....	1,285-1,310
Dull reddish shale and sandstone, soft	1,310-1,323
Reddish shale	1,323-1,338
Sandstone and dark shale	1,338-1,350
Dull reddish shale and a little sandstone....	1,350-1,391
Sandstone with layers of dark shale	1,391-1,435
Hard sandstone with soft streaks.	1,435-1,448
Sandstone and dark shale	1,448-1,461
Dark shale (thin streaks of lignite)	1,461-1,491
Light, hard shale..	1,491-1,531
Shale, not so hard ..	1,531-1,540
	1,540-1,566
Hard sandstone	1,566-1,576
Hard shale	1,576-1,601
Hard shale with soft streaks.	1,601-1,613
Hard shale ...	1,613-1,626
Very hard. Ironstone boulder.	1,626-1,633
Hard shale a little gas about 1,650 feet	1,633-1,682
Hard and soft shale alternating.	1,682-1,689
Shale and sandstone alternating	1,689-1,722
Shale with a little sand-stone	1,722-1,731
Shale, soft and dark ...	1,731-1,736
Hard sand-rock....	1,736-1,747
Shale...	1,747-1,752
Shale and sandstone	1,752-1,759
Shale....	1,759-1,763
Hard, supposed sandstone ..	1,763-1,767
Soft shale	1,767-1,770

[1]Dawson gives the following section at Athabaska, obtained from a natural exposure and from the bore-hole, the zero datum being the top of the bore-hole or about 1,660 feet above sea-level.

Height. Feet.	Thickness of Formation. Feet.
Top of bank.	
180—Yellowish sandstones, thin beds, with some iron-stone: *Foxhill* or *Laramie*.	15
165—Probably all grey shales, with some thin sandstone layers; not well exposed.	
Depth.	
0—Top of bore-hole.	
1,090—Grey and blackish shales, often very soft, with occasional thin, hard, layers of sandstone or iron stone.' Much gas at different levels between 245 feet and 780 feet: *LaBiche shales*	1,255
1,130—Grey sandstone, with a flow of soft water: *Pelican sandstone*....	40
1,233—Dark shales, often soft, a little sandstone: *Pelican shales*	103
1,461—Grey sandstones and grey, reddish, and blackish shales; the sandstone sometimes very hard and probably nodular, as in outcrop at Grand Rapids: *Grand Rapids sandstone*	228
1,770—Dark and light-grey shales, generally hard, with some sandstone layers, particularly towards the base: *Clearwater shales* ...	309 (or more)
Total.	1,950

[1] Geol. Surv., Can., vol. XII, p. 144

Near Pelican river the boring was carried to a depth of 821 feet 6 inches in 1897 and continued to a depth of 837 feet in 1898. The bituminous sands were met at a depth of 750 feet and were penetrated 87 feet. They were found to consist of soft sandstone saturated with bitumen. Strong flows of gas were struck at 750 and 773 feet, and at 820 feet a tremendous flow was struck, the roaring of which could be heard at a distance of 3 miles or more. The flow was so strong that no progress could be made in drilling, and work was abandoned until 1898, when it was thought the force of the gas would be decreased sufficiently to permit of further operations.

¹But though there was a seeming decrease of pressure, upon operations being resumed in 1898, " the seeming decrease was found to be in a great measure due to the closing up of the outlet at the bottom of the casing by an asphalt-like mixture, composed of maltha or petroleum tar and sand. In fact, when boring operations were resumed on June 17, the difficulty was found to be intensified by the accumulations of this asphalt-like maltha in the bottom of the bore.

" The rapid expansion of the gas produced a very low temperature, and this chilled and solidified the tar, or maltha, until it became as adhesive as wax. As the tools cut it loose the gas would carry it up through the bore, until from bottom to top it was almost one mass of sand and tar. The only way it could be extracted from the sand-pump was by heating the latter over a fire; even then very little could be got out at one time, it being so thick that it was almost impossible to force it up into the pump. I used different sorts of tools to cut it off the walls and clean it out, but the longer we worked at the bore the greater the quantity of tar accumulating on the sides of the casing and tools." By using smaller casing the hole was carried to a depth of 837 feet, when another flow of gas was met, nearly equal in volume to that met at 820 feet, and the work was stopped.

²A sample of the gas taken by F. H. McLearn was analysed by E. Stansfield, of the Mines Branch, Department of Mines, with results as follows:

CO_2	1·0 per cent.
O	2·9
Methane	83·5
N	12·6

Assuming oxygen present is due to contamination with air the composition would be

CO_2	1·2 per cent
Methane.	97·0
N	1·8

Log of Well near Pelican.³

	Feet.
Sand and gravel.	1-86
Very soft, dark-bluish shale.	86-101
Soft sandstone.	101-105
Very soft, dark-bluish shale. At 185 feet slightly saline water	105-185
Rather hard, reddish-brown shale	185-225
Sandstone. At 225 feet water	225-234
Sandstone and brown shale	234-245
Hard, grey shale. At 253 feet more water and gas	245-253
A light, greenish-grey shale	253-280
Soft, greenish-grey shale, cement like.	280-290
Brown shale, with strata of grey shale	290-308

¹ Geol. Surv., Can., vol. XI, p. 32A.
² Geol. Surv., Can., Sum. Rept., 1916, p. 151.
³ Geol. Surv., Can., Sum. Rept., 1897 and 1898.

Log of Well near Pelican—Concluded.

	Feet.
Brown shale	308-310
Hard sandstone. More gas and water..	310-311
Brown shale and sandstone in alternate strata	311-328
Sandstone	328-340
Brown shale	340-353
Hard sand-rock, with layers of softer rock (At 355 feet struck maltha and gas).	353-365
Sandstone, rather hard	365-410
Brown shale .	410-427
Hard, brown shale. . .	427-450
Sandstone. More gas and water. ...	450-465
Grey shale. .	465-526
Ironstone	526-532
Grey shale . .	532-553
Sandstone .. .	553-556
Very hard, probably ironstone..	556-558
Very hard sandstone .	558-563
Brown shale. .	563-573
Grey shale, streaks of sandstone.	573-590
Grey shale, brown shale and sandstone in alternating strata; the cuttings show traces of maltha	590-620
Grey shale. Strong flow of gas at 625 feet; considerable maltha coming away with the water.....	620-625
Very hard sandstone	625-643
Soft, grey shale	643-648
Hard sandstone	648-652
Soft, grey, sandy shale	652-665
Ironstone	665-675
Soft, grey shale...	675-684
Hard sandstone .	684-685
Soft, dark-grey shale... . . .	685-703
Hard sandstone.... .	703-713
Soft, grey, sandy shale	713-718
Hard sandstone	718-723
Sandstone........ .	723-733
Soft, grey shale... .	733-743
Soft, grey shale, with streaks of soft sandstone. Strong flow of gas at 750. A heavy oil mixed all through the sandstone and shale....	743-758
Soft, dark-grey shale, and soft sandstone. Heavy oil throughout. At 773, a heavier flow of gas........	758-781
Alternate strata of soft, grey shale and soft sandstone. Increased quantities of heavy petroleum. Gas increasing in volume.	781-800
Same as foregoing. At 820 feet, a tremendous flow of gas of which it car could be heard 3 miles or more........	800-820
Soft sandstone. Hard streak, and light flow of gas at .. a)	820-830
Soft sandstone. . . .	830-836
Iron-pyrites nodules embedded in cement-like sandstone. Very strong flow of gas...:	836-837

[1]Dr. Dawson gives the following section from this well:

Depth from surface. Feet.	Thickness of formation. Feet.
86—Sand and gravel (surface deposits).....	86
185—Dark, bluish-black, soft shales, with some sandstone in upper part . . . . *Pelican shales*......	99
465—Greyish sands and sandstones, and brownish and greyish shales. *Grand Rapids sandstones*...................	280
750—Greyish and brownish shales, alternating with thin beds of hard sandstone and ironstone. *Clearwater shales*.................	285
837—Sands and clays often saturated with heavy oils and tar. *Tar sands*................................	87 (or more)

[1] Geol. Surv., Can., vol. X, p. 19A.

A number of wells have been sunk along the Athabaska in recent years in search of petroleum, but the results have not been satisfactory.

The Pelican Oil and Gas Company, Limited, did some boring a few miles above Pelican river and in Well No. 3 of this company a strong flow of gas was struck.

Log of Well No. 4, Pelican Oil and Gas Company.[1]

	Feet.
Blue and yellow shale	1 - 66
White and grey shale (water)	66 - 82
Blue shale	82 - 200
Blue and brown shale	200 - 235
Brown shale	235 - 285
Grey-brown shale	285 - 331
Sand rock (hard)	331 - 352
Shale	352 - 365
Sand rock	365 - 425
Shale	425 - 507
Brown shell (hard)	507 - 509½
Grey shale	509½ - 538
Shell	538 - 540
Sandstone	540 - 546
Shale	540 - 575
Hard shell	575 - 581
Grey shale, streaks of sandstone	581 - 644
Strong flow gas	625
Grey shale (gas)	644 - 651
Grey shale (soft and vermont-like)	651 - 666
Sand rock	666 - 671
Grey shale	671 - 688
Hard, brown shell	688 - 689
Dark grey shale	689 - 740
Hard shell	740 - 741
Dark grey shale	741 - 766
Hard shell	766 - 767
Dark grey shale, sandy	767 - 843½
Sandy shale	843½ - 872
Sandy shale	872 - 882
Coarse rock mixed with heavy oil	882 - 887
Shale and sand	887 - 898
Hard rock	898 - 903
Lime carrying oil	903 - 997
Limestone	997 - 1,051
Hard, flinty shell (strong flow of gas under shell)	1,051 - 1,053½
Limestone	1,053½ - 1,158
Hard, fine shell	1,158 - 1,159
Limestone	1,159 - 1,192
Hard shell (gypsum)	1,192 - 1,197
Blue shale and gypsum	1,197 - 1,293
Hard, lime shell	1,293 - 1,296
Lime rock	1,296 - 1,538
Lime, shale, and lime rock	1,538 - 1,560
Grey shale and lime (gas)	1,560 - 1,700
Limestone	1,700 - 1,784
Hard shell	1,784 - 1,790
Lime rock (shale streaks)	1,790 - 1,875
Hard shell	1,875 - 1,879
Layers of limestone and shale	1,879 - 2,040
Strong flow of gas	2,040
Limestone and shale, interstratified	2,040 - 2,069

Several companies have carried on boring operations along the river from McMurray to some distance below McKay and in some of the wells small quantities of oil have been struck. The logs of two wells bored at the mouth of Horse creek in the vicinity of McMurray are given on page 128.

[1] L. G. Huntley, Am. Inst. Min. Eng., Bull. 102, p. 1550.

Log of Well No. 1, Athabaska Oils, Limited, Located 9 Miles below McKay.

	Feet.
Top soil....	1- 4
Cemented oil sand.	4- 18
Oil stratum..	18- 78
Shale.....	78- 98
Oil stratum..	98- 155
Shale.. ...	155- 330
Limestone...	330- 395
Shale. .	395- 415
Limestone . .	415- 330
Gypsum.	330 720
Limestone.	720- 912
Red rock...	912- 975
Hard, reddish flint sand or rock....	975 1105
Reddish quartz or rock.	1105 1130

Salt water was struck at depths of 765 and 1,000 feet. A small quantity of petroleum was struck in the upper part of the well.

During the summer of 1916 boring operations were conducted at the upper end of Vermilion chutes and at a point on Peace river 17 miles below Peace River Crossing.

At Vermilion chutes boring was carried to a depth of 860 feet through Devonian limestone and shale. The upper beds are porous and impregnated with bitumen. The well below Peace River Crossing was drilled to a depth of over 1,100 feet, and at 857 feet a small quantity of heavy black oil was struck. This was cased off and the hole was continued to a depth of over 1,100 feet when a very strong flow of gas was struck. An analysis of a sample of the oil is given in the Summary Report of the Geological Survey, Canada, for 1916.

F. H. McLearn in his report for 1917 states:

"The discovery of oil this summer in the No. 2 well of the Peace River Oil Company, about 15 miles below Peace River, lends particular interest to this part of the district. The well is situated on the high strata where the south dip flattens out northward. The oil occurs at two horizons, beds of sandstone in both cases, near the base of the Loon river and not far above the limestone contact. The upper sandstone is met with from 842 to 948 feet in the drill-hole. Above 852 feet this bed yielded gas; from 852 to 905 feet it contained a highly viscous oil; from 905 to 910 feet it carried salt water, and from 910 to 948 feet was firmly cemented and barren of oil, gas, or water. Below this is a 14-foot shale bed followed below from 962 to 1,032 feet by the second oil sand. This is impregnated with an oil of somewhat better quality. This horizon would produce a few barrels per day. An analysis by the Mines Branch (Edgar Stansfield, Chief Engineering chemist) gave:

(a) Specific gravity at 60°F.. 981
(b) Distillation test.

Below 150° C . .	2.0% by volume naphtha.
" 150-200 . .	4.87
" 200-250 .	5.3
" 250-300 .	56.2 66.3 % illuminating oils.
" 300-325 .	5.2
Residue and loss	26.5 % lubricating oil.

"Oil of similar gravity was found at the corresponding horizons in the No. 1 well 1½ miles downstream. The two sandstones are of less thickness here and the shale between thicker. The thinning of these oil sands northward no doubt limits the possibilities of exploration in that direction, since with their disappearance there would be no reservoir to contain oil. As far as structure alone is effective, it should be noted that it is of a gentle nature north of the wells, with very low dips. The conditions obtaining at the wells, therefore, might be expected to prevail over a considerable area from the wells north, limited in that direction more particularly by the wedging out of the sandstones. Were the strata thrown into undulations of a more pronounced nature, more circumscribed areas favourable to boring could be pointed out. In accordance with the recently stated theory of David White, it is not improbable that the heavy gravity of the oil is related to this gentle structure.

" To the west past Dunvegan and St. John, no favourable structures for boring were located. At Hudson Hope there is a broad low anticline. The anticline at the Gates is broken by a fault. About 600 feet below river-level the upper shales, sandstones, etc., of the Bull Head Mountain sandstone would be met with. Below this the lower sandstones of that formation are too massive to serve as a suitable reservoir. The Triassic shales and sandstones must be over 3,000 feet below river-level at Hudson Hope.

Log, No. 2 Well.[1]

Well No. 2, Peace River Oil Co., Peace River, Alberta.
Begun July 21, 1917.
Drilled to 980 feet, October 13, 1917.
Elevation above River-level on October 15, 1917 —32 feet. Water at 38 feet.

Material.	Thickness. Feet.	Depth. Feet.
Sand, earth, gravel	10	10
Coarse gravel	46	56
Blue clay	33	89
Dark blue shale—a little gas at 89 feet	36	125
Sand rock	19	144
Blue shale—thin bands sand rock. At 253 feet, small showing gas	162	306
Grey sand rock. At 308 feet, gas with faint showing of oil	23	329
Blue shale—very small showing salt water 327-329 feet.	54	383
Sand rock	12	395
Blue shale	62	457
Sand rock, gas at 459 feet—smells of oil	16	473
Blue shale	25	498
Lime rock	6	504
Blue shale	66	570
Sand rock	11	581
Blue shale	31	612
Lime rock	38	650
Blue shale	25	675
Lime rock	36	711
Blue shale	51	762
Lime rock	17	779
Blue shale	63	842
Sand rock. Hard at 842 feet; at 843 feet softer with gas; at 845 feet gas strong—strong smell oil; 845-848 feet heavy flow gas, sand soft; 848-852 feet, very hard; at 852 feet softer with slight amount oil; 878 feet more oil, deeper; from Friday night until Monday morning about 2 barrels oil in well; 905-910 feet soft sand rock, getting salt water; oil stops here; 934-944 feet very hard sand rock; no water here	106	948
Blue shale	14	962
Sandstone, with oil; fills to 120 feet in less than a day after bailing; hole 6-in. diameter at this depth.	18	980
Oil-bearing sandstone	52	1,032
Sandstone—carrying water above	11	1,043
Sand with thick tar	4	1,047
Sand	10	1,057

[1] Log furnished by the driller.

The possibilities of the Devonian sediments as a source of petroleum are worthy of consideration. These sediments cover a large portion of the Mackenzie basin, disappearing to the south beneath the Cretaceous rocks. In many places they are highly bituminous in character, and several petroleum springs are known.

The upper beds penetrated by the boring at Vermilion chutes are impregnated with bitumen, and strata exposed near La Butte on Slave river and in several places around the western arm of Great Slave lake are also highly bituminous. Bitumen occurs in limestone at the Ramparts, in some of the dark argillaceous shales exposed at the water's edge along Grand View, and in shales in a small plateau lying a short distance east of the Mackenzie 15 miles below Grand View. The shales in this plateau are reddened in places by combustion of the bituminous matter. Fifteen miles farther down the river evenly-bedded, highly bituminous, black, Devonian shales occur. "The laminæ, when freshly separated, are moistened on the surface with an oily liquid, and burn when thrown into the fire, and patches of red shales, marking the sites of former fires, alternate with the dark varieties. The shales are exposed in the right bank for some miles, or almost as far as old Fort Good Hope. They dip down the river at a low angle."[1]

Petroleum springs issue from the Devonian rocks at a number of points. Probably the best known are those that occur at Windy point on the north shore of Great Slave lake.

"The springs are situated a couple of hundred yards from the shore, at the base of a low limestone cliff, which runs inland from the lake, and are three in number, each of them being surrounded with a small basin, three to four feet in diameter, filled with inspissated bitumen, while the soil and moss for some distance away are impregnated with the same material. A small quantity of pitch is annually taken from these springs and used for boat building purposes, while a much larger supply could be obtained if needed. A sulphur spring resembling ".. .e at Sulphur point on the south shore of the lake, but much more   issues from the foot of the cliff in close proximity to the bituminou   ..s, and feeds a considerable stream.

"The rock through which the petroleum ascends here is a heavily bedded, greyish rather coarsely crystalline, cavernous dolomite, and is entirely unlike the bituminous beds south of the lake and down the Mackenzie, which in most cases consist of calcareous shales. The dolomite is everywhere permeated with bituminous matter, which collects in the numerous cavities, and oozing up through the cracks, often forms small pools on the surface of the rock.

"The age of the bituminous beds here could not be clearly ascertained, as they are entirely unfossiliferous, but it is altogether likely that they are older than the Devonian shales and limestone which outcrop along the southern shore, and are more nearly related to the dolomites which underlie the fossiliferous Devonian beds at Nahanni butte and at other places. The presence of bitumen in such abundance here also suggests an anticlinal which would bring up lower beds.

"Sulphur and tar springs are reported to occur at a point about halfway between this and Fort Rae, but as I did not hear of them until I had

[1] McConnell, R. G., Geol. Surv., Can., Ann. Rept., vol. IV, p. 110D.

left the lake, I was unable to visit them. A tar spring is also known to exist under the surface of the water in the deep bay immediately east of the Big Island fishery, as many of the boulders and rocks along the shore in this neighbourhood ar coated with bitumen which has been washed ashore, and hummocks of ice stained with the same material are often observed. On the south shore bituminous shales and limestones occur at several points, and it would thus appear that the oil-bearing beds underlie the whole western part of the lake."[1]

A range of hills composed of dolomite rises from the shore on the north side of Sulphur bay which lies to the north of Windy point. Tar springs comparable in size to those of Windy point are found on these hills at an elevation of 200 feet above the lake and a mile inland.[2]

Some of the limestone in the "Rock-by-the-river-side" is slightly bituminous and Franklin observed "sulphurous springs and streams of mineral pitch issuing from the lower parts of the limestone strata" exposed in Bear rock.

McConnell states that "near Fort Good Hope several tar springs exist, and it is from these that the Hudson's Bay Company now obtain their principal supply of pitch. The springs are situated at some distance from the river and were not examined."[3]

From the point of view of the prospector for oil or gas the McMurray sandstone is so far as present knowledge goes the most important formation of Cretaceous age in the Mackenzie basin. It comprises the great body of bituminous sands exposed on Athabaska river and is the formation from which issue the strong flows of natural gas struck in the wells sunk by the Dominion Government and by the Pelican Oil and Gas Company in the vicinity of Pelican river. In the government well the sands were found to be bituminous between a depth of 750 and 837 feet, where boring ceased. Although strong flows of gas were obtained the results of the operation were disappointing in that petroleum was not obtained. The non-occurrence, however, of petroleum at this point, is no proof that it does not occur at other points where the McMurray sandstone is overlain by impervious strata.

Our knowledge of the structure of the Cretaceous formations in the Athabaska River district is very limited, but such information as we have may be of use to those anticipating further operations. The Cretaceous sediments of this district are nearly horizontal, and rest upon the Devonian with an unconformity so slight that it can be detected only by examination of the contact for a distance of several miles. Between Grand rapids and Pelican rapids, the dip of the Cretaceous strata is about $5\frac{1}{2}$ feet a mile south, and between Pelican rapids and Athabaska the dip is about 10 feet a mile up the river. This greater dip between Pelican rapids and Athabaska may be due, in part, to its southwest direction and may indicate that the true dip of the Cretaceous strata on the Athabaska is west of south. A low anticline crosses the river near Crooked rapids; the dip on each side is only 3 or 4 feet per mile. Below McMurray the strata are probably nearly horizontal and may have a slight north or northwest dip.

[1] McConnell, R. G., Geol. Surv., Can., Ann. Rept., vol. IV, 1891, p. 75D.
[2] Cameron, A. E., Geol. Surv., Can., Sum. Rept., 1916.
[3] Geol Surv, Can., Ann. Rept., vol. IV, p. 31D, 1891.

To test the McMurray sandstone where it is overlain by impervious strata it will, therefore, be necessary to conduct boring operations to the south or southwest of its outcrop.

The oil struck in the shallow wells of the Athabaska Oils, Limited, opposite the mouth of Dover river, occurs in a hollow of the Cretaceous-Devonian unconformity. This depression is 12 miles long in the direction of the river and opposite the mouth of Dover river has a depth of about 140 feet below the limestone rim. Further downstream there is a similar smaller depression.[1]

The problem of the origin of the hydrocarbons of the McMurray sandstone has not been solved. One theory is that they were derived from underlying bituminous sediments of Devonian age; another, that they were forced into the porous sandstone from the sediments of later deposition.[2]

The structure of the Cretaceous on Peace river is described by F. H. McLearn as follows:

" The salient features of the foothill structure in this district are the long, low, east dips, of 10 degrees and less, and the steep west dips. The latter, effecting only 2 miles or less of section, and separated by 10 to 12 miles of low east dip, are probably related to incipient overthrusting. The structure is a whole represents the dying out of the effects of the Rocky Mountain overthrust. The border of the foothills is marked by an anticlinal structure pitching southward. This consists of a single, large anticline on Portage mountain and in the canyon and of two anticlines on Bull Head mountain.

" The transition from foothills to plains structure is very abrupt and takes place where the Portage Mountain anticline is succeeded by an area of gentle undulation and overthrust faulting extending as far east as the Gates. This includes a broad, low anticline at Hudson Hope and a small broken anticline at the Gates. From there to Cache creek there is a low east dip, under one-half degree, with a local west dip equally low near the mouth of Cache creek. Eastward to several miles below St. John the structure seems to be almost flat. Near the North Pine river and downstream the structure steepens with an east dip so as to bring the Dunvegan sandstone almost to river-level a few miles below the mouth of Kiskatinaw river. From here to the bend at Montagneuse river the structure is flat. The section is east-west to this point. From the mouth of Montagneuse river southward a north-south section is cut and here a small south dip is revealed. Where the river turns east past Dunvegan an east-west section is again exposed and flat structure indicated. Beyond the mouth of Burnt river, the Peace turns to the northeast and so continues to Peace river. Here the strata rise downstream and the inclination near Peace river amounts to some 40 feet per mile to the south.

" From Peace river northward a north-south section is exposed and at first reveals a south dip of about 10 feet per mile. In the vicinity of the No. 2 well and extending to Tar island the structure is practically flat, although there is probably a slight rise of 1 or 2 feet per mile. Downstream from here there is a slight dip north of a few feet per mile to a point about 10 miles below the mouth of Cadotte river. Beyond this there is a gentle rise, and a final flattening out. The above structure applies to

[1] McLearn. F. H., Geol. Surv., Can., Sum. Rept., 1916.
[2] Huntley, L. G., Am. Inst., Min. Eng., Bull. 102, pp. 1345-1348.

that observed above river-level. It is possible, however, owing to the thinning of the Loon River shales northward and the consequent rise of the limestone contact, that the lower strata below river-level, which would be reached by drilling, would be slightly tilted southward as compared with the overlying strata above river-level. This applies particularly to the section north of Tar island."

The Devonian strata exposed on Athabaska, Peace, and Mackenzie rivers are nearly horizontal. On Athabaska river and its tributaries there is some local doming, and limestone is brought to the surface by an anticline between 35 and 40 miles up Mikkwa river, a tributary of the Peace.

A. E. Cameron, in his report for 1917, describes the structure of the Devonian system on Great Slave lake is as follows:

"Stratigraphic studies have revealed the presence of a gentle anticline stretching across the lake from Pine point on the south shore to Nintsi (Windy) point on the north shore. On the south shore the apex of the anticline is shown on the east side of Pine point and the lowest beds exposed show the Pine Point series of bituminous limestones and limy shales. The presence of higher strata, the overlying Presqu'ile dolomites, both to the east on Burnt islands and to the west at Presqu'ile point, clearly demonstrates the anticlinal structure. On the north shore the crest appears to lie in the erosion basin of Sulphur bay. The Presqu'ile dolomites are exposed on the east shore of Nintsi point, at various points on the shores of the bay, and eastward to Jones point; whereas the overlying Slave Point limestones are shown on Slave point west of Nintsi point and again near House and Moraine points northeast of Jones point.

"Structural conditions suggestive of gentle anticlinal folding are noticeable in the limestone outcrops on Buffalo river and those exposed on Hay river above the falls. Exposures are confined to the valley floors when the rivers have cut down into the Devonian sediments and on account of the overburden of glacial drift the extent of the folding was not observable.

"As shown by the oil seepages and tar pools on Nintsi point the Presqu'ile dolomites appear as the most probable oil horizon of the district and as these sediments are exposed on the limits of the anticline, the possibilities of an oil producing field existing on the shores of the lake are not very great.

"A thick series of soft clay shales is exposed on Hay river, which from the stratigraphical relation would appear to overlie the dolomites. If the dolomite horizon can be found elsewhere in the district under suitable structural conditions and overlain by these impervious shales, it may be worth investigating with a drill.

"On Hay river, above the falls, limestone outcrops in the valley show gentle undulations forming anticlines and synclines of a low order. The limestones here exposed represent upper members of the Hay River limestones which overlie the thick series of shales above mentioned.

"It is to be noted that the section exposed on Peace river shows members of the Simpson shale series unconformably overlying the gypsum series of upper Silurian age, and thus the middle Devonian section is exposed on Great Slave lake is here absent. It is, therefore, possible that the dolomites would not be found underlying the shale series in the folded area above the falls on Hay river, or, if present, they would probably be somewhat thinner in their development than is shown on the shores of the lake. If drilling operations were conducted on Hay river, a thickness of about 1,000 feet of sediments would have to be pierced before the Presqu'ile dolomites would be reached."

A few borings have been made into or through the Devonian limestones and shales, but no petroleum has been struck. Below McMurray a number of wells have been sunk on local domes; at McMurray and 9 miles below McKay, wells have penetrated to the Pre-Cambrian rocks, and a short distance above Pelican river on the Athabaska a boring was carried through the Cretaceous and 1,000 feet into the Devonian. In the last well a flow of gas was encountered at a depth of 2,040 feet. At Vermilion chutes a well was sunk to a depth of 860 feet through Devonian strata. The few borings that have been made have not given encouraging results, but the field for exploration is exceedingly broad.

POTASH.

Considerable interest has been manifested in the possible occurrence of potash in commercial quantities in the Mackenzie basin. The analyses of samples taken from brine springs and from waters rising from strata pierced in boring operations, as given on pages 129 and 130, are disappointing in this regard. The proportion of potassium is too small to be of commercial im: "nce or to be indicative of the presence of beds of potassium minerals. Analyses of a sample of water rising from a depth of 268 feet, in a bore-hole put down at Vermilion chutes on Peace river, and of a sample taken from a natural spring at Sulphur point on the south shore of Great Slave lake, are equally disappointing.[1]

SALT.

Numerous brine springs are found in the Mackenzie basin, from some of which salt is obtained for local consumption.

It is reported that rock salt or salt-bearing formation was encountered in Devonian strata in two wells sunk by the Northern Alberta Exploration Company on Horse creek near McMurray. As the wells are only 155 feet apart there is a strange lack of agreement between the logs.[2]

	Well No. 1	Well No. 2
	Feet.	Feet.
Loose surface material	0- 17	0- 24
Limestone	17- 117	24- 124
Soapstone and limestone	117- 520	124- 590
Salt	520- 620	590- 690
Limestone	620- 635	690- 765
Salt	635- 740	765- 855
Limestone	740- 770	855- 935
Sandstone	770-1,475	935-1,406

[1] Camsell, Charles, Geol. Surv., Can., Sum. Rept., 1916.
[2] Mines Branch, Dept. of Mines, Can., "Report on the salt deposits of Canada and the salt industry," 1915, p. 84.

¹The springs that are economically the most important at the present day are those found at the base of the escarpment lying to the west and south of the forks of Salt river west of Slave river (Plate XIV). Although there are here four groups of springs, from each of which salt is deposited, there are only two fro which it is being collected at present. About four tons of salt is collected annually for the use of the trading posts and missions of the Mackenzie River district.

The water of the springs flows into shallow circular basins, from which it trickles away through barren, salt-encrusted clay flats to the river. The basins are usually 15 or 20 feet in diameter and the bottoms are covered with salt deposited through the evaporation of the brine. In some cases hillocks of salt 12 or 15 feet in diameter and as much as 2 feet in height are formed at the springs.

At the time of Camsell's visit to these springs on August 26 and 27, 1916, the flow was rarely as much as 4 gallons per minute at any one spring, but is said to be considerably greater in the spring season.

Samples of brine taken from three localities were analysed by the Mines Branch, Department of Mines, Ottawa.

No. 1 locality, Hudson Bay springs at the forks of Salt river; taken August 21, 1916.

Contains in 1,000 parts by weight:

Ions.		Hypothetical combination.	
Potassium	0·5	Potassium chloride	0·9
Sodium	134·5	Sodium chloride	258·0
Calcium	1·2	Calcium sulphate	4·1
Magnesium	0·2	Sodium sulphate	0·4
Chlorine	157·7	Magnesium chloride	0·8
Sulphuric acid (SO₄)	3·1		
	264·2		264·2

Temperature of air on collection . . . . 62° F.
Temperature of brine . . . . 40° F.
Specific gravity at 65°F. . . . . 1·204
Flow about 1½ gallons per minute from each of eight springs.

No. 2 locality, Mission springs, about 6 miles south of the forks of Salt river; taken August 26, 1916.

Contains in 1,000 parts by weight.

Ions.		Hypothetical combination.	
Potassium	0·4	Potassium chloride	0·8
Sodium	100·8	Sodium chloride	256·3
Calcium	1·2	Calcium sulphate	4·2
Magnesium	0·2	Sodium sulphate	0·2
Chlorine	156·6	Magnesium chloride	0·8
Sulphuric acid (SO₄)	3·1		
	262·3		262·3

Temperature of air on collection . . . . 70° F.
Temperature of brine . . . . 35° F.
Specific gravity at 65°F. . . . . 1·204
Flow about 3 gallons per minute.

¹ Camsell, Charles, Geol. Surv., Can., Sum. Rept., 1916.

No. 3 locality, Snake Mountain springs, about 2 miles east of Mission spring; taken August 29, 1916.

Contains in 1,000 parts by weight:

Ions.		Hypothetical combination.	
Potassium.................................	·4	Potassium chloride..	0·8
Sodium....................................	10·7	Sodium chloride......	256·0
Calcium...................................	00·2	Calcium sulphate	4·2
Magnesium................................	0·2	Sodium sulphate. .	0·2
Chlorine..................................	156·4	Magnesium chloride...	0·8
Sulphuric acid (SO₄)....	3·1		
	262·0		262·0

Temperature of air on collection.........	55° F.
Temperature of brine...........	40° F.
Specific gravity at 65°F.	1·202
Flow about 4 to 5 gallons per minute.	

In recalculating these analyses we find that sodium chloride constitutes in each sample over 97·6 per cent of the total solids. The percentage of dissolved matter in the brine, namely over 26 per cent, indicates practically a saturated solution of salt at that temperature.

The salt is probably derived from crystals of salt disseminated through the gypsum that is found in abundance in the escarpment. Its presence in at least some of the gypsum is shown by an efflorescence of common salt that appears on some specimens when exposed to the atmosphere. Salt-bearing gypsum of this nature occurs at Snake Mountain springs:

At La Saline, on Athabaska river, 28 miles below the mouth of the Clearwater, there are several mineral springs about half a mile east of the river on the edge of the valley. The deposits, which consist principally of calcareous tufa carrying a small amount of common salt, gypsum, and native sulphur, cover the face of the escarpment and have built up a cone 10 to 15 feet high and about 200 feet wide. Sulphuretted hydrogen gas escapes from the bank in several places. The water from the springs flows into a shallow lake, which is situated at the foot of the escarpment, and is surrounded by a clay flat partly bare and partly covered with coarse grasses.[1] Following are the results of the analysis of a sample of water collected from this spring by Mr. McConnell.

[2]"Specific gravity, at 15·5°C., 1·052.

[1] McConnell, R. G., Geol. Surv., Can., Ann. Rept., vol. V, p. 35D.
[2] Hoffmann, G. C., Geol. Surv., Can., Ann. Rept., vol. VI, p. 79R.

"Agreeably with the results of an analysis conducted by Mr. Wait, it contained, in 1,000 parts, by weight:

Potassium	0·868
Sodium	23·937
Calcium	1·574
Magnesium	0·496
Sulphuric acid (SO₄)	4·702
Chlorine	38·461
	70·038
Chlorine required, in addition to that found to satisfy bases	0·056
	70·094

"Hypothetical combination:

Chloride of potassium	1·655
" sodium	60·883
" magnesium	1·049
Sulphate of lime	5·352
" magnesia	1·155
	70·094
Total dissolved solid matter, by direct experiment, dried at 180°C.	69·616

"There was not enough of the water at the disposal of the operator to admit of his examining it for any of the more rarely occurring constituents."

A copious saline spring bubbles up 100 feet from the west bank of Athabaska river, 2 miles above the mouth of P'delay creek, and feeds a considerable stream. Large quantities of sulphuretted hydrogen gas are given off[1]. An analysis of a sample of the water collected by Mr. McConnell gave the following results:

[2]"Specific gravity at 15·5°C., 1·012. An analysis, by Mr. Wait, showed it to contain, in 1,000 parts, by weight:

Potassium	0·036
Sodium	4·783
Calcium	0·947
Magnesium	0·122
Sulphuric acid (SO₄)	2·759
Chlorine	7·394
	16·041
Chlorine required, in addition to that found, to satisfy bases	0·021
	16·062

"Hypothetical combination:

Chloride of potassium	0·069
" sodium	12·165
Sulphate of lime	3·220
" magnesia	0·608
	16·062
Total dissolved solid matter, by direct experiment, dried at 180°C	16·263

"The quantity of water at the disposal of the operator was too limited to allow of his examining it for any of the more rarely occurring constituents."

[1] McConnell, R. G., Geol. Surv., Can., Ann. Rept., vol. V, p. 36D.
[2] Hoffmann, G. C., Geol. Surv., Can., Ann. Rept., vol. VI, p. 80R.

Samples of brines were collected by S. C. Ells in the McMurray district as follows and analysed.

No. 1. Overflow from casing-head of No. 1 well, Athabaska Oils, Ltd., Athabaska river.
No. 2. Overflow from casing-head of "Salt of the Earth" well, and. by A. von Hammerstein, on west bank Athabaska river, 1 mile north of McKay.
No. 3. From largest spring at La Saline lake.
No. 4. Overflow from casing of well drilled by Fort McKay Oil and Asphalt Company, at La Saline (August, 1914).

Results of Analyses of Brines.[1]

	No. 1	No. 2	No. 3	No. 4
	Parts per million.	Parts per million	Parts per million.	Parts per million.
Ca	1668	1747	1821	3351
Mg.	385	585	571	1021
K	296	346	496	192
Na.	22988	70268	21184	84076
HCO₃	469	372	530	36
CO₃.	none	none	none	none
Cl.	36188	118636	39792	127960
SO₄.	4144	4920	4888	2956
Sp. gr. at 15·5 C.	1·047	1·133	1·052	1·150

"A saline spring, emitting natural gas and carrying up small quantities of tar, occurs on the boulder beach at the upper end of Tar island" in Peace river about 30 miles below the mouth of Smoky river, and a second spring is reported to occur on an island opposite the mouth of White Mud river.[2]

A long saline slough, 60 feet wide, occurs in sec. 25, tp. 106, 5th mer., and a saline river, 125 feet wide and with a slight current, crosses sec. 1, tp. 107.[3] Several springs of salt water, accompanied by sulphuretted hydrogen, bubble up close to the left bank of Christina river at a point about 12 miles up from its mouth.[4]

Numerous mineral springs occur along Clearwater river. The most notable group occurs on the north side about 4 miles below Cascade rapid. "Here the springs are very copious, issuing from the bank in a number of places, for a space of 300 yards in length. The largest single spring forms a small brook itself, and the addition of these and all the other mineral springs which flow in farther down, must increase considerably the soluble salts in the water of the whole river."[5] Five and a half quarts of water was taken from one spring and evaporated. A residue of 1·36 ounces (avoir-dupois) of crude salt was obtained and from one-fifth to one-fourth more adhered to the kettle used in evaporating. The salt consisted of very

[1] Mines Branch, Dept. of Mines, Canada, Sum. Rept., 1914, p. 64.
[2] McConnell, R. G., Geol. Surv., Can., Ann. Rept., vol. V, p. 49D.
[3] Ponton, A. W., Dept. of the Interior, Top. Surv. Branch, Can., Ann. Rept., 1908-1909, p. 171.
[4] McConnell, R. G., Geol. Surv., Can., Ann. Rept., vol. V, p. 39D.
[5] Bell, Robt., Geol. Surv. Can., Rept. of Prog., 1882-83-84, p. 26CC.

large quantities of soda, chlorine, and sulphuric acid, large quantities of lime, magnesia, and carbonic acid, a small quantitiy of insoluble residue, and very small quantities of potash, alumina, and ferric oxide.

Richardson states that a little below the rapid on Great Bear river "a small stream from the southward flows into the Bear Lake river, near whose sources the Indians procure an excellent common salt, which is deposited from the springs by spontaneous evaporation."[2] He also states that a thermal spring, much resembling sea-water in its saline contents, issues from the front of the cliff known as the "Rock-by-the-river-side", and the fissure from which it flows is incrusted with crystallized gypsum. An analysis of the water showed that the chief salt was sulphate of magnesia.[3] Richardson was informed by Mr. McPherson that a salt spring, having a basin 15 feet in diameter, which is never dry, occurs on the top of Nahanni butte.[4] This butte stands at the junction of the Nahanni and Liard rivers. Mr. McConnell, who ascended it in 1887, did not find the spring, but says "a neighbouring mountain, however, showed a white patch on its steep side which is plainly due to the deposits of a mineral spring of some kind, and may be the one referred to."[5]

According to Petitot, salt occurs on mount Clark, and in this mountain two saline rivers, known as Big and Little Salt rivers, take their rise.[6]

An outcrop of rock salt occurs on the northeast end of Bear rock and salt from this outcrop has been used by the mission at Norman. Rock salt has also been reported at Norman as occurring in the mountains to the west.

SILVER.

The report that silver ore similar to the Cobalt ore had been discovered in the vicinity of Fond du Lac at the east end of lake Athabaska led to a great influx of prospectors into this region in 1915. The district was visited by Charles Camsell in the summer of 1915, but an examination of the prospects between Grease river and Camille bay failed to reveal any deposits of silver.[7] A brief description of the geology is given on pages 113 to 115.

A specimen weighing half an ounce and consisting of "an association of grey mica-schist with a white subtranslucent quartz more or less thickly coated with hydrated peroxide of iron, carrying some coarsely crystalline galena" was assayed for Mr. E. Lyon, who reported having procured it from the south side of Great Slave lake about 40 miles east of Resolution. It was found to contain:

Gold............none.
Silver......... 16·012 ounces to the ton of 2,000 pounds.

"The galena amounted to 41·2 per cent, by weight, of the whole; hence the same, freed from all gangue, would contain at the rate of 38·865 ounces of silver per ton of 2,000 pounds."[8]

[1] Adams, F. D., Geol. Surv., Can., Rept. of Prog., 1882-83-84, p. 18MM.
[2] "Narrative of a second expedition to the shores of the polar sea in the years 1825, 1826, and 1827", by John Franklin. Appendix No. 1, p. 13.
[3] "Arctic searching expedition: a journal of a boat-voyage through Rupert's Land and the Arctic sea in search of the discovery ships under command of Sir John Franklin." vol. I, p. 181.
[4] "Arctic searching expedition: a journal of a boat-voyage through Rupert's Land and the Arctic sea in search of the discovery ships under command of Sir John Franklin." vol. II, p. 203.
[5] Geol. Surv., Can., Ann. Rept., vol. IV, p. 57D.
[6] Bulletin de la Société de Géographie, 6th ser., vol. X, 1875, p. 162.
[7] Camsell, Charles, Geol. Surv., Can., Sum. Rept., 1915, p. 122.
[8] Hoffmann, G. C., Geol. Surv., Can., Ann. Rept., vol. XI, p. 33R.

A. Typical topography of the Laurentian plateau: near Chipewyan (Pages 11, 12, 13.)

B. Reindeer mountains, east of the Mackenzie delta (Page 22.)

· PLATE III.

A. Athabaska river near Grand rapids. (Page 22.)

B. Grand rapids, on Athabaska river. (Page 23.)

A. Lower falls on Hay river: height 46 feet. (Page 29.)

B. Falls on Tazin river.

138

PLATE V.

Garden at Fort Smith. (Page 41.)

PLATE VI.

Hay meadow on the salt plain west of Fort Smith. (Page 42.)

PLATE VII.

Buffalo range southwest of Fort Smith. (Page 47.)

PLATE VIII.

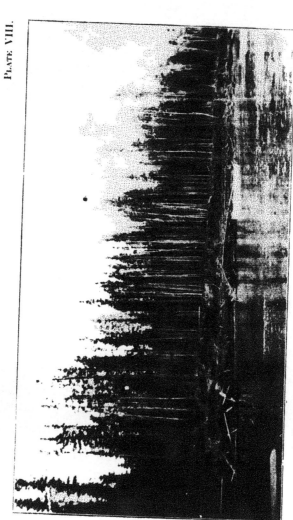

Forested country on Peace river below Little rapids. (Page 48.)

A. Chipewyan. Photograph by H. J. Bury, Dept. of Indian Affairs. (Page 56.)

B. Fort McPherson. (Page 56.)

PLATE X.

Top of the Devonian limestone below McMurray on Athabaska river. (Pages 64 and 70.)

PLATE XI.

A. Gorge between Alexandra falls and Lower falls on Hay river, looking down stream. (Page 65.)

B. The Ramparts, Mackenzie river. Photograph by H. J. Bury, Dept. of Indian Affairs. (Page 69.)

A. Exposure of bituminous sand on the north side of Steepbank river, 21 miles from its mouth, illustrating typical massive structure and cleavage of many of the high-grade deposits. (Pages 71 and 92.)

B. Exposure on the north side of Steepbank river, 2 miles from its mouth, showing low-grade and banded bituminous sand, overlying Devonian limestone (Pages 71 and 92.)

146

PLATE XIII.

Cliff of gypsum below Peace point, Peace river.   (Page 111)

PLATE XIV.

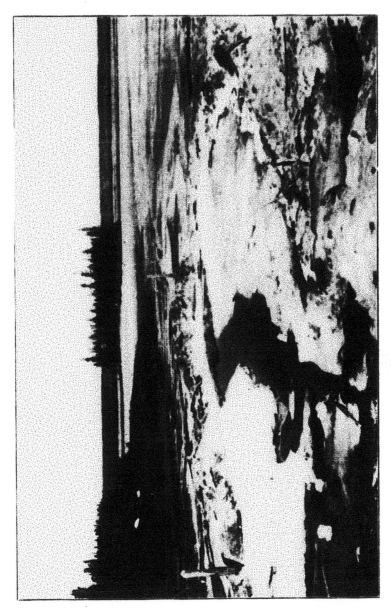

Mission salt springs, southwest of Fort Smith. (Page 150.)

INDEX.

L.

M.

N.

O.

P.

CANADA

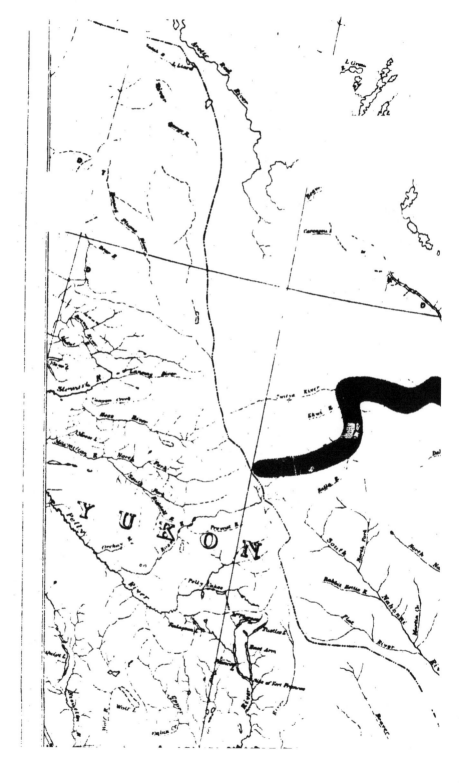

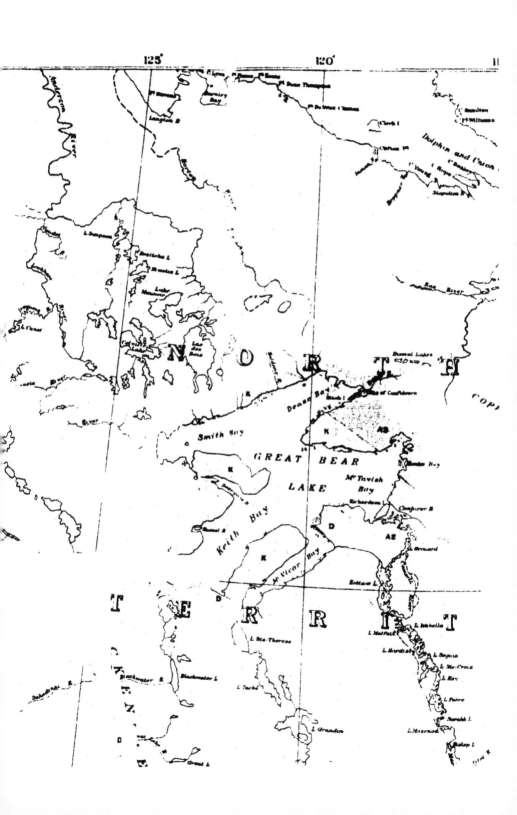

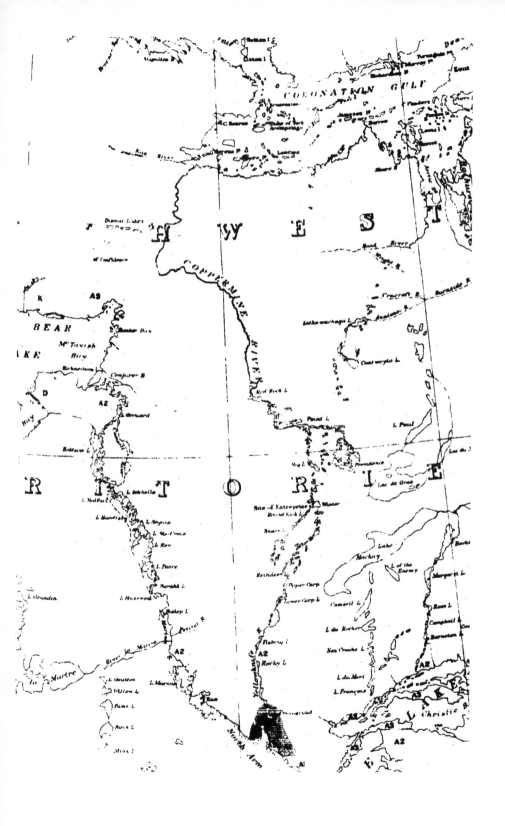

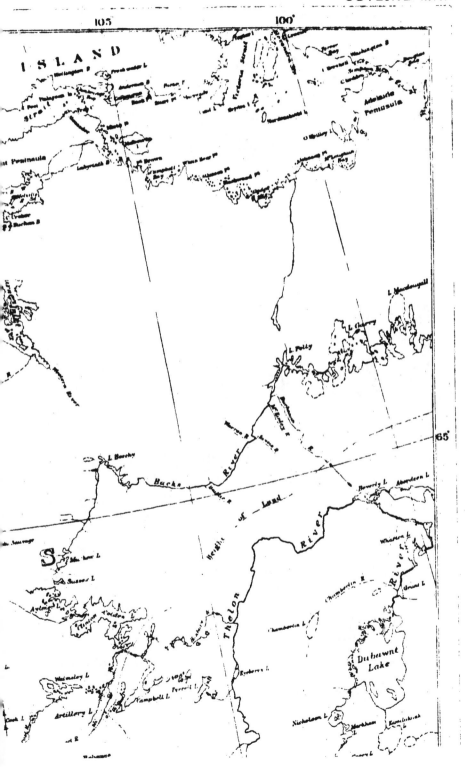

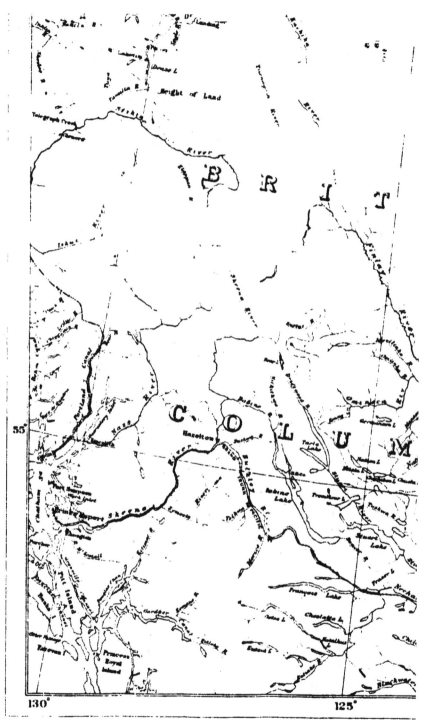

C.O.Senécal, *Geographer and Chief Draughtsman*
J.O.Fortin, *Draughtsman*

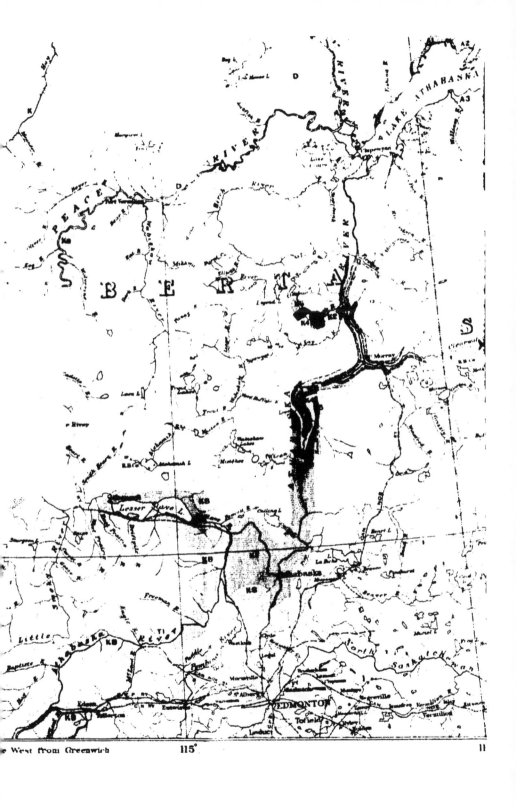

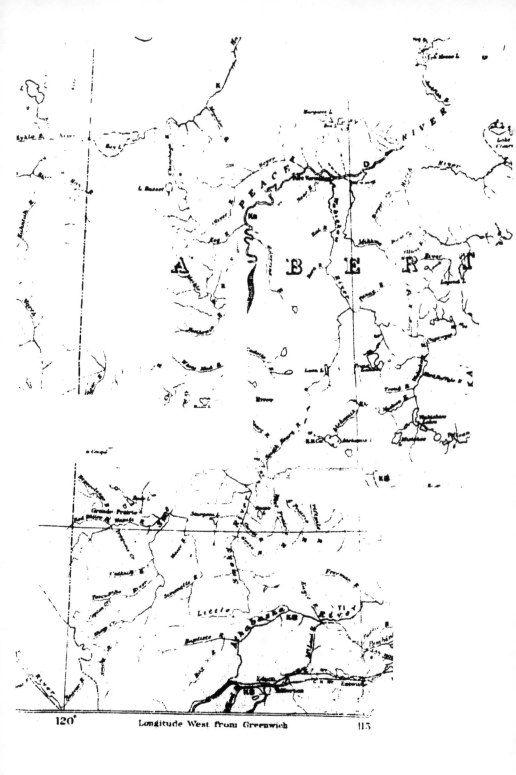

MICROCOPY RESOLUTION TEST CHART

(ANSI and ISO TEST CHART No. 2)

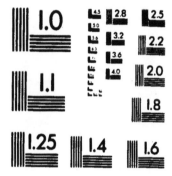

APPLIED IMAGE Inc

1653 East Main Street
Rochester, New York   14609   USA
(716) 482 - 0300 - Phone
(716) 288 - 5989 - Fax·